CAN A VIRUS CAUSE SCHIZOPHRENIA?
Facts and Hypotheses

NEUROBIOLOGICAL FOUNDATION OF ABERRANT BEHAVIORS

CAN A VIRUS CAUSE SCHIZOPHRENIA?
Facts and Hypotheses

by

Bradley D. Pearce
Emory University School of Medicine, Atlanta, Georgia

KLUWER ACADEMIC PUBLISHERS
Boston / Dordrecht / New York / London

Distributors for North, Central and South America:
Kluwer Academic Publishers
101 Philip Drive
Assinippi Park
Norwell, Massachusetts 02061 USA
Telephone (781) 871-6600
Fax (781) 681-9045
E-Mail: kluwer@wkap.com

Distributors for all other countries:
Kluwer Academic Publishers Group
Post Office Box 322
3300 AH Dordrecht, THE NETHERLANDS
Telephone 31 786 576 000
Fax 31 786 576 474
E-Mail: services@wkap.nl

Electronic Services <http://www.wkap.nl>

Library of Congress Cataloging-in-Publication Data

Title: CAN A VIRUS CAUSE SCHIZOPHRENIA? Facts and Hypotheses
Author: Pearce, Bradley D.
ISBN: 1-4020-7300-3

A C.I.P. Catalogue record for this book is available
from the Library of Congress.

Printed on acid-free paper.

Printed in the United States of America.

This book is dedicated to my parents, Mary Jane and Robert Pearce, for their patience and loving support.

Contents

Preface

The purpose of this book is to provide a critical appraisal of the concept that schizophrenia can be caused by viral infections. The ability of viruses to cause psychiatric *symptoms* is not in question – the mental depression following a bout with mononucleosis or hallucinations associated with herpes encephalitis are well-described examples. However, a viral origin for chronic disorders such as autism, schizophrenia, and bipolar disorder, is another matter. The claim of an infectious etiology for these disorders has been met in the mainstream scientific community by a vague skepticism that occasionally erupts into stringent criticism. Too often, however, the viral hypotheses of these disorders is simply disregarded; marginalized with the hope that it will go away so that the "serious" research aimed at uncovering the real cause of these illnesses will not lose its focus. This being said, much of the criticism is valid, and as a researcher with formal training in both neuroscience and viral immunology, I view the proposed viral etiology of these disorders with a skeptical eye.

Nevertheless, one cannot ignore the growing number of well-performed studies pointing to the role of viral infections as important antecedents of schizophrenia and other disorders in the schizophrenia spectrum. In the last 30 years there have been hundreds of articles in peer-reviewed journals presenting evidence or positing theories to suggest that at least some cases of schizophrenia have a viral origin. Moreover, many schizophrenia experts have been calling for the recognition that schizophrenia is a heterogeneous group of disorders that may have different causes. This idea of disease heterogeneity is reaching a crescendo, and there is undoubtedly a place for viruses among alternative etiologies; but we have to look. The intellectual climate to look now is better than it was 20 years ago, in part because in the last 2 decades a number of chronic diseases of unknown etiology (e.g. gastric ulcers, Kaposi's sarcoma, hepatocellular carcinoma) have been ascribed to infectious causes.

In this volume I endeavor to assemble the pertinent schizophrenia literature into a set of testable hypotheses which incorporate the latest findings in neurobiology, virology, and developmental neuroscience. There is a growing consensus that both genetic and environmental factors are involved in schizophrenia. Implicit in most current theories of schizophrenia pathogenesis is the occurrence of a neurodevelopmental insult that precedes the symptoms by decades. This has lead to the pessimistic suggestion that victims of schizophrenia may be "doomed from the womb." The idea that the culprit is a virus has far reaching consequences, suggesting the more optimistic possibility that we could "vaccinate to eliminate" this scourge of the human mind. Indeed, if only 10% of schizophrenia cases were found to be engendered by preventable infections, this would still hold the potential to eliminate a tremendous amount of the personal grief and medical costs that this disease imposes on society. Moreover, we are currently in an exciting new era in the development of effective antiviral medications, and this opens a glimmer of hope for treating at least one underlying cause of schizophrenia. Interestingly, some antipsychotic medications have been shown to have anti-viral properties.

This book is intended primarily to meet the needs of researchers and clinicians who are interested in the etiology and pathogenesis of schizophrenia. Although I anticipate that psychiatrists, psychologists and neurobiologists will comprise the bulk of the readership, I have structured the book such that most sections will also be of value to researchers and graduate students who have a tangential interest in neuropsychiatric disorders but maintain their primary focus in neurovirology, immunology, neuropathology or perinatology. Because my goal for this volume is to integrate across disciplines and marshal the disparate data and theories concerning schizophrenia into a mechanistic framework, I have attempted to make the content accessible to researchers who may not be specialists in the diverse fields represented. In this regard, chapter two includes an introduction to virology and immunology. In addition, I have strived to disseminate sufficient introductory material throughout the volume so as to allow the non-specialist to grasp the salient concepts put forth.

Acknowledgements

I wish to acknowledge Kristina Weis for her assistance in preparing this book and assembling the tables. I also thank Dr. Melissa Demetrikopoulos for her critical reading of the manuscript.

Chapter 1

INTRODUCTION AND HISTORY

1.1 A BRIEF HISTORY OF THE VIRAL HYPOTHESIS

The idea that schizophrenia is caused by a contagious agent is rooted in beliefs that are intertwined with the early conceptions of the disease. Long before the various symptoms of schizophrenia were consolidated nosologically into a definable disease entity (first as dementia praecox and later as schizophrenia), madness (insanity) was posited to have an infectious cause. Aristotle noted the contagious nature of madness in dogs, writing in a History of Animals "... that dogs suffer from the madness. This causes them to become very irritable and all animals they bite become diseased" (Nathanson 1997). Aristotle was, of course, referring to rabies but he seemed to miss the connection between being bit by a dog and the development of rabies in humans. Even so, statements acknowledging the infectious nature of rabies from dog to human have been traced back as far as the 23rd century BC, where reference is made to rabies in the Eshmuna Code of Babylon (Kaprowski et al. 1997). Rabies is one of the most dreaded afflictions of human kind. The fear of an inevitable and agonizing death is heightened by the dramatic changes in behavior that accompany the furious form of the disease. Neuropsychiatric manifestations of rabies include agitation, distorted thinking, and sometimes hallucinations and delusions. Occasionally, a person with rabies will act aggressively or violently. For example, in the 1950's a California house painter suffering from furious rabies bit 20 people in one night (Kaprowski & Dietzschold 1997).

Although the rabies virus can undoubtedly cause an acute form of insanity with a latency (incubation period) that can range from days to several years, the usual symptoms in humans, and certainly the disease course, bear little resemblance to schizophrenia. Nevertheless, as the definition of schizophrenia evolved in the early 1900's, the concept that other infectious agents were involved was considered quite seriously. Indeed, both Kraeplin and Bleuler entertained the idea of an infectious cause for schizophrenia, and Mennninger was convinced that at least some cases could be attributed to an infectious encephalitis (Torrey 1988).

When placed in the historical context of the first few decades of the 20th century, these seemingly portentous views of schizophrenia pathogenesis are not surprising. Dementia paralytica, which had hitherto been thought to be caused by prodigious alcohol consumption and sexual depravity, had just recently been found to be due to an infectious agent, syphilis. There was a growing realization that humans have always co-existed with a rich and diverse microbial sub-world, and Koch had put forth his famous postulates defining the causative relationship between a disease and an infectious pathogen. What's more, within the first three decades of the 1900's the nascent field of virology, spawned by the discovery of filterable infectious agents, had found several human and animal diseases attributable to this minute and elusive microbe (Nathanson 1997).

In addition, the pandemic of viral encephalitis lethargica occurring between 1919 of and 1928 drew attention to the capability of infectious agents to induce complex neurobehavioral sequela. Several authors noted the similarity between psychotic features experienced by patients following a bout with encephalitis lethargica, and the clinical presentation of schizophrenia (Lishman 1987). Similarly, post influenza psychosis was well recognized during this period. In this regard, Menninger'sMenniinger's study of psychotic symptoms following influenza led him to formulate the first clear viral hypothesis of schizophrenia (Menninger 1926).

During the first third of the 20th century at least some cases of serious mental illness were widely regarded to be underlain by a brain disorder, and the proposition of an infectious cause was by no means anathema. Nevertheless, this relatively enlightened view was to be eclipsed in the second third of that century as Freudian psychology took firm hold, and psychodynamic/social interpretations of schizophrenia held sway until the late 1950's. Moreover, the belief that schizophrenia was associated with definable neuropathological alterations was being challenged. This was exemplified initially in the mid 1920's by histological studies that could find no changes in schizophrenic brains compared to matched controls. This period of negative neuropathological findings culminated in 1972 with Plum's dictum that "schizophrenia is the graveyard of neuropathologists." (Plum 1972)

In fairness, this opposition was well-grounded in the scientific knowledge of the day. To put this into perspective, one needs to consider that more than a century has passed since the first pathological examination of schizophrenia brain tissue, yet we find ourselves at the beginning of the 21st century unable to point to a single distinctive, consistent, and reproducible neurohistological abnormality that characterizes the disease. Nevertheless, few doubt that schizophrenia is a brain disease, and technical refinements in molecular pathology and brain imaging have permitted a convincing case to be made that there are subtle yet relatively common abnormalities in brain regions such as the hippocampus and pre-frontal cortex. Moreover, as discussed further in Chapter 4, a confluence of data indicates that schizophrenia may have its origins in neurodevelopment.

In the late 1970's there began a resurgence of interest in the viral hypothesis of schizophrenia. This was propelled, to a large extent by developments in virology itself. Between 1950 and 1970, technical refinements in multiple fields led to a new era in virology. Prior to 1950, even quantifying the amount of an animal virus in a clinical or experimental isolate could be a difficult and imprecise undertaking. With the refinement of the plaque assay for quantifying infectious virus, and the use of the electron microscope to visualize virion structure, the number and diversity of animal viruses began to be fully realized (Nathanson 1997). Viral immunology was also becoming more sophisticated during this period, and new tools arose which facilitated the identification of viral subtypes and revealed specific mechanisms of pathogenesis. And while virologists had always been at the forefront of molecular genetics, the landmark accomplishment of deciphering DNA's structure and defining its central role in eukaryotic inheritance profoundly accelerated the pace of discovery in virology and microbial pathogenesis.

These developments conspired to vastly broaden the understanding of potential mechanisms by which viruses could interact with mamallian hosts. Even so, there was only a smattering of interest in the possibility of a viral etiopathogenesis for schizophrenia in the 1950's and 1960's (Lilenfeld et al. 1955; Morozov 1954; Taft et al. 1964). This began to change in the mid 1970's following a publication by E. Fuller Torrey and Michael Peterson that reconsidered the role of viruses in schizophrenia (Torrey et al. 1973). A subsequent paper further marshaled the evidence supporting the viral hypothesis (Torrey et al. 1976). These articles were inspired by the discovery of the curious properties of a "slow virus" that caused Kuru--a neurodegenerative disorder that is now known to be caused by prions. Torrey, an original thinker and gifted orator, has continued to be a leading proponent of the viral hypothesis for schizophrenia. Over the last three decades, he has effectively dampened the enthusiasm of his critics by his rigorous application of the scientific method, and the diversity of approaches he has taken.

In the late 1970's and early 1980's a number of other investigators joined the fray and put forth a myriad of creative theories, each of which seemed to hold some promise of revealing the viral connection with schizophrenia (Crow 1978; Dwyer 1979; Gajdusek 1978; Hare 1979; Knight 1984; Lycke et al. 1974). The most prolific of these has been Robert Yolken, who as director of the Stanley Neurovirology Laboratories has pressed the envelope toward more advanced and sophisticated investigations of the viral hypotheses (Johnston-Wilson et al. 2001). Epidemiological investigations, spearheaded by Mednick and colleagues, have also become more sophisticated (Brown et al. 2002; Mednick et al. 1994). And the ability to study Borna Disease virus and influenza in animal models, as well as the recognition that diverse pathogens can potentially share pathological mechanisms linked through common immune responses or neurotransmitter changes, has further advanced our basic understanding of the disease processes that may underlie schizophrenia. However, at present there is no single string of evidence that can be followed to a definitive judgment on the viral hypotheses. Instead, we must assess the hypotheses in its current state: an accretion of evidence derived from disparate sources and viewpoints that forms an irregular mass that is unsatisfying to look at, but no less compelling.

REFERENCES

Brown AS, Susser ES. In utero infection and adult schizophrenia. Mental Retardation & Developmental Disabilities Research Reviews 2002; 8:51-7.

Crow TJ. Viral causes of psychiatric disease. Postgraduate Medical Journal 1978; 54:763-7.

Dwyer DS. Virus-like particles in C.S.F. in schizophrenia [letter]. Lancet 1979; 1:1184-5.

Gajdusek DC. The possible role of slow virus infection in chronic schizophrenic dementia. Birth Defects: Original Article Series 1978; 14:81-7.

Hare EH. Schizophrenia as an infectious disease. British Journal of Psychiatry 1979; 135:468-70.

Johnston-Wilson NL, Bouton CM, Pevsner J et al. Emerging technologies for large-scale screening of human tissues and fluids in the study of severe psychiatric disease. International Journal of Neuropsychopharmacology 2001; 4:83-92.

Kaprowski H, Dietzschold B.Rabies: lessons from the past and glimpes of the future. In: PK Peterson; JS Remington, editors In defense of the brain: current concepts in the immunopathogenesis and clinical aspects of CNS infections. Malden: Blackwell; 1997.

Knight JG. Is schizophrenia an autoimmune disease? A review. Methods & Findings in Experimental & Clinical Pharmacology 1984; 6:395-403.

Lilenfeld AM, Pasamanick B, Rogers M. Relationship between pregnancy experience and the development of certain neuropsychiatric disorders in childhood. Am. J. Pub. Health 1955; 45:636-43.

Lishman WA. Organic Psychiatry. Oxford: Blackwell; 1987.

Lycke E, Norrby R, Roos BE. A serological study on mentally ill patients with particular reference to the prevalence of herpes virus infections. British Journal of Psychiatry 1974; 124:273-9.

Mednick SA, Huttunen MO, Machon RA. Prenatal influenza infections and adult schizophrenia. Schizophrenia Bulletin 1994; 20:263-7.

Menninger KA. Influenza and schizophrenia. An analysis of post-influenzal "dementia praecox" as of 1918, and five years later. American Journal of Psychiatry 1926; 5:469-529.

Morozov MA. On the problem of the virus etiology of schizophrenia. Korsakov J. Neuropathology and Psychiatry 1954; 54:732-4.

Nathanson N.Introduction and history. In: N Nathanson, editor Viral Pathogenesis. Philadelphia: Lippincott-Raven; 1997; p. 3-11.

Plum F. Prospects for research on schizophrenia. Nueropathological findings. Neuroscience Research Program Bulletin 1972; 10:384-8.

Taft LT, Goldfarb W. Prenatal and perinatal factors in childhood schizophrenia. Dev. Med. and Child Neurol. 1964; 6:32.

Torrey EF. Stalking the schizovirus. Schizophrenia Bulletin 1988; 14:223-9.

Torrey EF, Peterson MR. Slow and latent viruses in schizophrenia. Lancet 1973; 2:22-4.

Torrey EF, Peterson MR. The viral hypothesis of schizophrenia. Schizophrenia Bulletin 1976; 2:136-46.

Chapter 2

FUNDAMENTALS OF VIRAL INFECTIONS AND IMMUNE RESPONSES IN HUMANS

2.1 A PRIMER ON GENERAL VIROLOGY

In contrast to bacteria and fungi, viruses are obligate intracellular parasites that depend on host cells to replicate their genome. From the perspective of a cell, viruses are like pirates, latching on and boarding the ship (cell), enslaving its crew (metabolic and biosynthetic machinery), and frequently scuttling the vessel (cell lysis or death) when it no longer suites the invader's requirements. From the virus's perspective, however, infectivity is essential in order to have an enduring existence in nature. Accordingly, viruses have evolved diverse strategies for replication within living cells, and have found niches as intracellular parasites of plants, animals and unicellular organisms.

Viruses come in a plethora of sizes and shapes (figure 1), but functionally they can be envisioned to consist of a genetic payload encapsulated in a delivery system (Dermody et al. 2000). A schematic of a virus particle, also called a virion, is depicted in the figure 2. The payload of the virus consists of the viral genome, along with the core proteins that are needed to commence the early stages of the virus's replication cycle before it gains control of the host cellular machinery. The outer (delivery vehicle) portion of the virion serves as a barrier against degradation, and additionally displays proteins that facilitate viral attachment to the cell exterior. It consists of a capsid shell, which is made up of an orderly arrangement of protein subunits called capsomers. Many viruses are also enveloped where by they have an outermost membrane that is derived from the host cell during budding-- a process that usually occurs near the end of the virus replication cycle.

Once a virion is formed and exits the cell, it must go through several steps in order to propagate itself in another susceptible cell (figure 3). First it must become attached or adsorbed to the outside of the host cell membrane. If a virus is to reproduce, it must find a route to the cell interior.

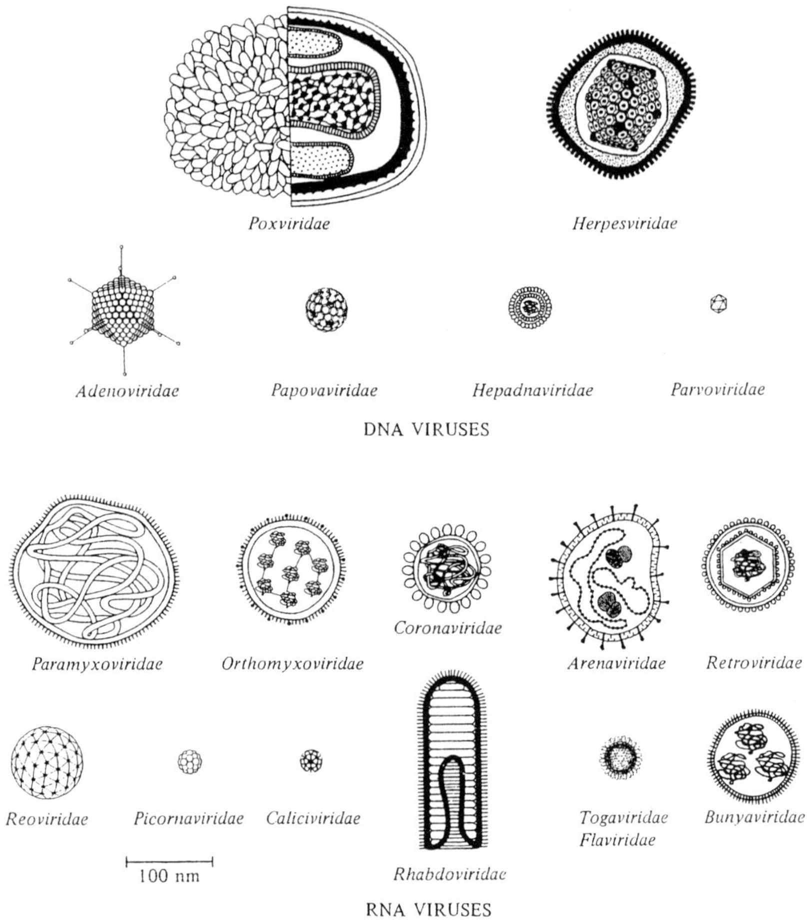

Figure 1. Diagram illustrating the shapes and sizes of animal viruses of families that include human pathogens. The virions are drawn to scale, but artistic license has been used in representing their structure. In some, the cross-sectional structures of capsid and envelope are shown, with a diagrammatic representation of the genome; with the very small virions, only their size and symmetry; in the largest, both. Reproduced with permission from: Fenner, F. & White, D., ed. (1986) *Medical virology, 3rd edition* (pp. 24). Orlando: Academic Press, Inc.

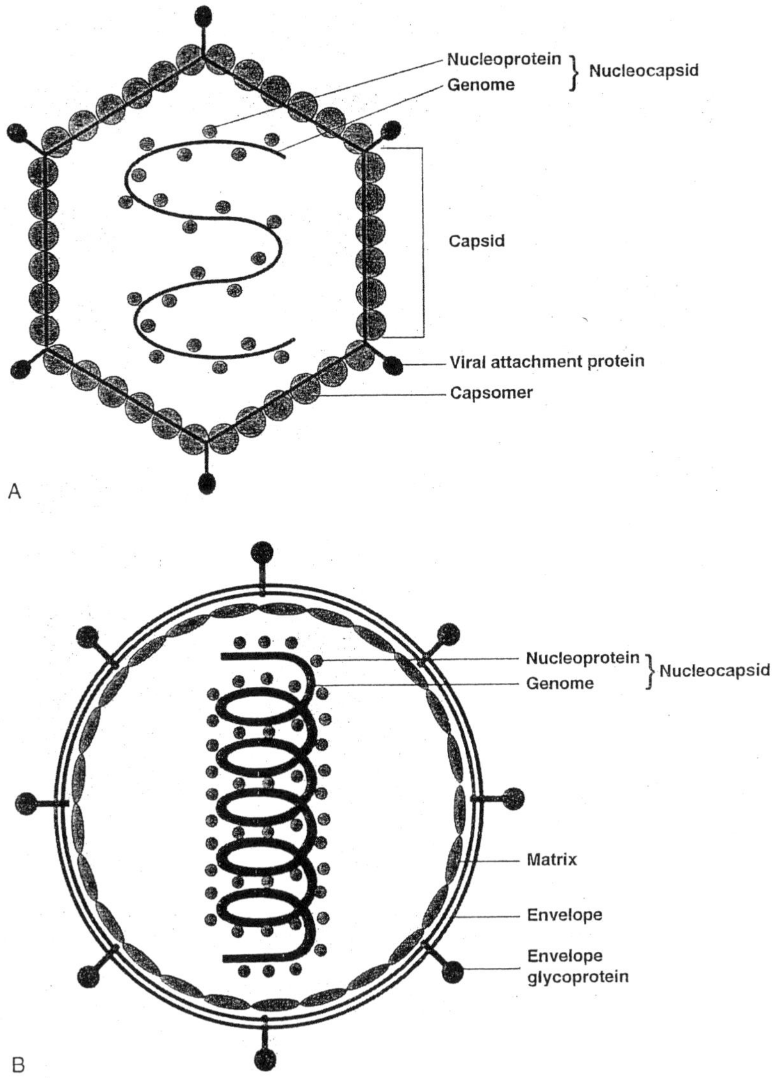

Figure 2. Schematic diagrams of the structure of a novenveloped icosahedral virus (A) and an enveloped helical virus (B). Reproduced with permission from: Mandell, G.L., Bennett, J.E., & Dolin, R., ed. (2000) Mandell, Douglas, and Bennett's principles of infectious diseases, 5th edition. Philadelphia: Churchill Livingstone. (Originally prepared by Mehmet Goral, Vanderbilt University, Nashville, Tenn.)

Therefore, viruses have evolved specific proteins on their surfaces that engage receptors on susceptible cells, allowing viral entry, fusion, and uncoating so that the genetic payload may be delivered inside the cell. Some of the eukaryotic surface receptors allowing the selective entry of viruses have been characterized, though many more remain to be discovered (table 1).

Because of the potential for genetic polymorphisms in these host surface receptors, these molecules represent one of the inherited variables conferring individual differences in susceptibility to viral infections. For example, CD4 and CCR5 serve as co-receptors for the entry of HIV into T-cells and macrophages. However, individuals who are homozygous for an uncommon deletion mutation of CCR5 show dramatically reduced rates of HIV infection despite extensive exposure to the virus (Samson et al. 1996). While this example illustrates a polymorphism conferring resistance to infection, the converse can also occur. Depending on the virus, a variant allele could code for a receptor that makes an individual more susceptible to infection. In a sense this is a tautology because the existence of a polymorphism encoding increased susceptibility is meaningfully defined only in contradistinction to an alternative allele conferring decreased susceptibility. Still, an important area of the viral epidemiology is to relate patterns of the viral receptor alleles with morbidity data in a population. With regard to viruses and psychiatric disorders, one can posit infection occurring only in a subset of the population who carry this susceptibility gene.

Once a virus has entered a cell and liberated its genome, the next step is the transcription and translation of its genetic instructions. Virologists have deciphered an impressive array of strategies that viruses employ to achieve this step (Brooks et al. 2001; Dermody & Tyler 2000). However, in simple terms the goal of the virus at this stage is to replicate it's genome and present the cell with mRNA transcripts that can be translated into the proteins needed for assembly of new virions (figure 4). Once the viral progeny have been assembled, replete with their newly synthesized DNA or RNA genomic core, they exit the cell and perpetuate their pernicious sojourn.

Table 1 Putative viral receptors.

Family	Virus	Possible receptor
Adenoviridae	Adenovirus type 2	Integrin $a_v\beta_3$ and $a_v\beta_5$
Coronaviridae	MHV	Carcinoembryonic antigens
	TGEV, 229E	Aminopeptidase N
	OC43, bovine	Sialic acid residues
Herpesviridae	EBV	CD21 (CR2 receptor)
	HSV	Heparan sulfate
	CMV	Heparan sulfate
		β_2Microglobulin/MHC I
	BHV-1, BHV-4	Heparan sulfate
	Pseudorabies	Heparin sulfate
Myxoviridae	Influenza A, B	Sialic acid residues
	Sendai	Gangliosides
	Measles	CD46
		Moesin
Parvoviridae	B19	Erythrocyte P antigen (globoside)
Picornaviridae	Polio	IgG superfamily (PVR)
	Rhinoviruses (major group)	ICAM-1
	Echovirus 1, 8	Integrin VLA-2 ($a_2\beta_1$)
Poxviridae	Vaccinia	EGF receptor
Reoviridae	Reovirus T3 (Dearing)	Sialic acid residues
		EGF receptor
	Rotavirus SA11	Gangliosides
		Sialic acids
Retroviridae	HIV-1	CD4
		Galactosylceramide
	MuLV	y+ amino acid transporter
	GLV	Phosphate permease
Rhabdoviridae	Rabies	Acetylcholine receptor
		Gangliosides
		Phospholipids
	VSV	Phosphatidyl serine
Togaviridae	Sindbis	Laminin receptor
	Semliki Forest	HLA H2-K, H2-D
	Lactate dehydrogenase	Ia

Some putative receptors listed remain the subject of ongoing debate and controversy. BHV, bovine herpes virus; EGF, epidermal growth factor; GLV, Gross leukemia virus; IgG, immunoglobulin G; MHC, major histocompatibility complex; MHV, mouse hepatitis virus; MuLV, murine leukemia virus; PVR, poliovirus receptor; TGEV, transmissible gastroenteritis virus; VSV, vesicular stomatitis virus. Reproduced with permission from: Fields, B., ed. (1996) *Fields Virology* (pp. 187). Philadelphia:, Lippincott Williams & Wilkins.

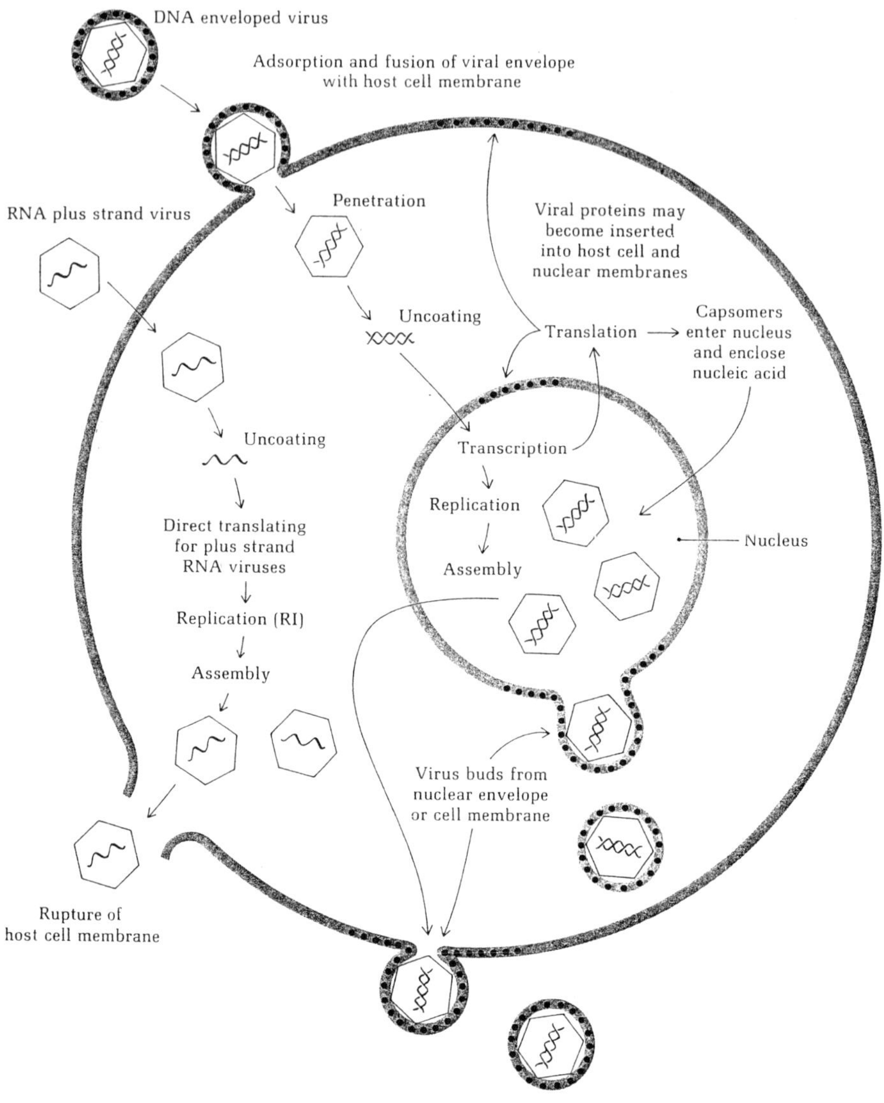

Figure 3. Schematic drawing illustrating the replication of an enveloped DNA virus and an RNA plus strand virus. Note, however, that with some DNA viruses, such as smallpox virus, the replication occurs solely in the cytoplasm of the host cell. In this drawing of viral replication, the outer membrane of the nuclear envelope has been omitted for clarity. Reproduced with permission from: Volk, W.A. & Wheeler, M.F. (1980) *Basic microbiology, 4th edition* (pp. 144). Philadelphia: J.B. Lippincott Company.

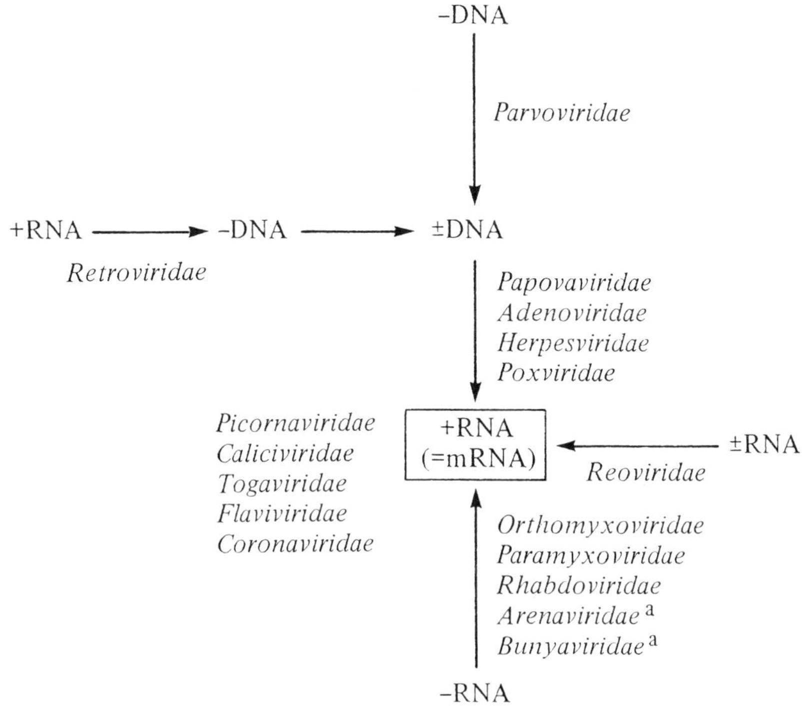

Figure 4. Six basic strategies for transcribing messenger RNA from different types of viral genome. Based on a scheme originally conceived by D. Baltimore, *Bacteriol. Rev.* **35**, 235 (1971). [a] At least some gene segments of the segmented RNA genomes of at least some arenaviruses and of the ***Phlebovirus*** genus of the ***Bunyaviridae*** family are ambisense, i.e., part +, part - polarity. Reproduced with permission from: Fenner, F. & White, D., ed. (1986) *Medical virology, 2nd edition* (pp. 63). Orlando: Academic Press, Inc.

2.1.1 Virus Nomenclature and Taxonomy

Viruses are named and classified according to size, morphology, genomic characteristics, replication strategy, organ tropisms, pathogenicity, presence or absence of an outer membrane, serological reactivity, or other physiochemical properties. It would be fair to argue that viral nomenclature is arbitrary or even capricious. However, virus taxonomy has improved, and viruses are currently grouped into orders, families, genera, and species, based on shared characteristics. Thus, a type1 poliovirus is formally termed Picornavirdae, Enterovirus, polio type 1. But in the clinical literature, one often encounters viruses referred to by their common name, for example poliovirus, or measles virus.

The most recent report from the International Committee on Taxonomy of Viruses is 1162 pages long (Van Regenmortel et al. 2000). It describes 3 orders, 240 genera, and 1550 species. For practical purposes this may be an underestimate of virus diversity; there are considerably more than 1550 distinct viruses in existence. In fact, the definition of a virus species is ambiguous, and within a given species there may be numerous strains with markedly different pathogenic properties. Importantly, the high rate of natural mutations that occur during the replication of many viruses, particularly those with RNA genomes, coupled with the ability of viruses to recombine or exchange their genetic material with other viruses, leads to continuous emergence of new distinct viruses. Granted, viruses capable of infecting humans are in the minority, but they still number in the hundreds. Since the early 1980's, an average of about 3 new viruses capable of infecting humans have been discovered each year (Khabbaz 1999). Whether this trend shows any sign of abating is still undecided, but it could even accelerate because of better virus detection methods, increased population mobility, and the expansion of human habitats. Thus, viruses represent a continuous yet dynamic threat to human health. Unfortunately, many of the mechanisms by which viruses cause disease, especially those characterized by subtle or delayed manifestations, have not been elucidated.

2.1.2 Disease Mechanisms

This section will provide a general overview of established mechanisms by which viruses cause human disease. Subsequent chapters will delve more deeply into specific mechanisms as they relate to psychiatric disorders.

Like most mammals, humans intermingle with viruses daily. Viruses can be transmitted through inhalation of aerosols, ingestion of contaminated food or water, hand contact with mucous membranes, sexual intercourse, insect bites, contaminated blood products, or breaks in the integument (Brooks et al. 2001; Dermody & Tyler 2000). Common viral transmission routes are shown diagrammatically in figure 5. A mechanism of transmission that is particularly germane to schizophrenia,is: the vertical transmission from mother to child. This can occur as a consequence of virus crossing the placenta or being secreted in the birth canal or breast milk.

On the average, virus exposure rarely results in overt disease. Often the virus is unable to breach the host integument and therefore it is unable to multiply. And even when infection does occur, symptoms of sufficient severity to come to clinical attention are the exception rather than the rule. Nevertheless, even asymptomatic infections are capable of altering cellular function and stimulating immune responses. Moreover, such infections may set into motion pathophysiological events that lie dormant for months or years before producing disease. The potential importance of such latent infections in schizophrenia will be discussed in more detail later on.

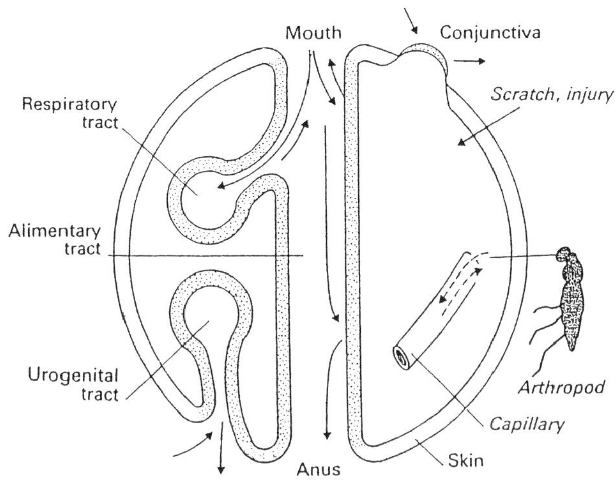

Figure 5. Body surfaces as sites of virus infection and shedding. Reproduced with permission from: Scheld, W.M., Whitley, R.J., & Durack, D.T., ed. (1997) *Infections of the central nervous system, 2nd edition* (pp. 11). Philadelphia: Lippincott-Raven Publishers.

The closely related terms "pathogenesis" and "virulence" are not interchangeable in the description of viral disease processes (Tyler et al. 1996). "Pathogenesis" refers to the mechanisms by which viruses interact with the host to cause disease. Accordingly, pathogenesis often concerns the specific mechanisms by which a virus injures cells or disrupts host homeostasis. Host factors play a central role in pathogenesis, and the impact of a virus is often secondary to the immune or inflammatory response. The term "virulence" is somewhat more oriented to the virus itself. Virulence is defined as the capacity of the virus to cause disease in a susceptible host. Hence, "virulence determinants" are viral molecules that confer the ability of one virus to cause disease compared with a closely related virus. Nevertheless, the virulence of a virus is dependent upon the host, and a virus may be highly virulent in one species but produce only asymptomatic infection in another. For example, "B virus" is harbored asymptomatically in rhesus monkeys but is usually fatal in the rare instances when it becomes contracted by a human.

The clinical outcome following a viral infection is determined by numerous interacting factors including the site of entry, dose of the inoculum, and "tissue tropisms" for the virus (Tyler & Fields 1996). Most viruses can infect only a subset of cell types within a susceptible host. The discreet cell types that permit replication of a given virus determine its tropism. Like many of the variables discussed so far, tropism is a function of both the virus and the host. An obvious determinant of tropism is the match between viral attachment proteins and host surface receptors permitting virus ingress. But this is not sufficient to allow viral replication because once a virus has entered the cell it must still deploy the requisite chemical and genetic signals to takeover the cellular machinery and control the events required to assemble infectious progeny. Thus viruses are often highly adapted to their host cells. This has even taken the form of the virus supplying temporary "life support" to the cell in order to buy enough time for the production of new virions. In this regard, several herpes viruses and human papilloma virus produce proteins that prevent apoptosis and consequently restrain the cell from its intended suicidal fate (Fazakerley et al. 2001).

2.1.3 Virus Spread

As shown in figure 6, viruses can spread and disseminate in the body by numerous routes. Often viruses travel hematogenously from their primary site of replication to distant cells or organs (Dermody & Tyler 2000). The virulence of a given virus may be manifested at secondary sites of replication. Poliovirus, for example, initially infects cells of the gastrointestinal tract but the disastrous consequences of polio are exacted from the host at the secondary site of the replication, the motor neurons of the spinal cord. Poliovirus causes cell lysis at both its primary and secondary sites of replication, but it is the inability of neurons to regenerate that results in permanent neurological disability.

Understandably, viral mechanisms in neuropsychiatric diseases usually invoke infection that attacks the mature or developing brain causing neuronal dysfunction that ultimately gives rise to behavioral abnormalities. It seems only reasonable that a prerequisite to disruption of neuronal structure or function is the ability of the culpable pathogen to actually enter the brain. Nevertheless, a virus could theoretically induce symptoms of schizophrenia or major depression without entering the CNS, for example, by stimulating antibodies that cross-react with brain tissue. Even so, given our current state of knowledge concerning viral mechanisms in psychiatric diseases, it's particularly important to understand the means by which viruses gain entry to the brain and cause perturbation of neural cells. Subsequent chapters will provide a more speculative treatment of this subject. For the purposes of this chapter, a general overview of known principles is in order.

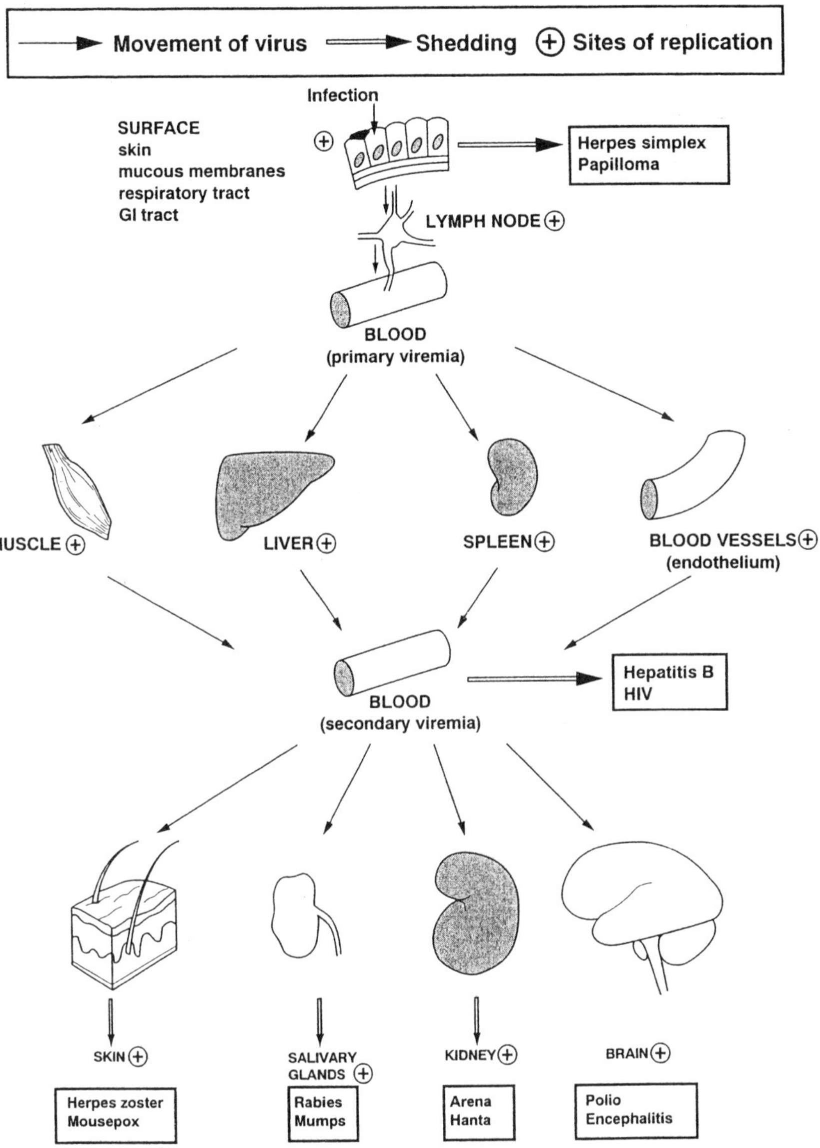

Figure 6. The life cycle of the virus in the infected host includes entry, dissemination, and shedding of blood-borne viruses. This generalized scheme emphasizes the sequential sites of and steps in virus replication: at the point of entry, in regional lymph nodes, production of primary viremia, replication in other tissues (different tissues for different viruses), production of secondary viremia, and invasion of target organs (different ones for different viruses), often leading to pathologic lesions and clinical disease. This scheme

applies to many but not all viruses and does not illustrate neural spread. Only a few examples are listed. GI, gastrointestinal; HIV, human immunodeficiency virus. Reproduced with permission from: Nathanson, N., ed. (1997) *Viral pathogenesis* (pp.15). Philadelphia: Lippincott-Raven Publishers. (Originally adapted from Mims CA. The pathogenesis of infections disease. London: Academic Press, 1987.)

2.1.4 Neural Infections

Various species and strains of viruses differ in their neuroinvasiveness (the ability to gain access to the CNS), neurotropism (ability to infect neural cells), and neurovirulence (ability to cause disease in the nervous system). As a first line of defense, the blood brain barrier deters the entry of microbes. This may help explain why most active infections with blood borne neurotropic viruses do not result in encephalitis. Nonetheless, viruses can access the CNS through several routes (figure 7). Defining the exact route in human infections is experimentally constrained because the spatio-temporal pattern of virus spread must be tracked from the initial portal of entry to the final destination in susceptible neural cells. Still, studies in experimental animals, along with the limited data in humans, have demonstrated the capability of viruses to exploit a variety of avenues to breach the protected milieu of the brain parenchyma (Johnson 1998). HIV uses monocyte/macrophages as a Trojan horse; the infected immune cells transit through the blood brain barrier carrying the virus with them. Some viruses cause cell-to-cell fusion resulting in a large dysfunctional syncytia further disseminating virus in the brain. Other viruses (Rhabdoviruses, Herpesviruses) use neurons as a "covered walkway", protected from the host defenses as they spread to distant regions of the CNS. This property has been used by neuroscientists to trace neuronal connections in a manner similar to the way a chemical tracer would be used, with the added advantage of a self-amplifying signal.

Considering the consequences of successful viral attack on the brain, it is not surprising that hosts have developed specialized protective mechanisms such as the blood brain barrier. The brain is distinguished from most peripheral organs by its unique collection of highly specialized differentiated cell types that are interconnected extensively and (for most neurons at least) are relatively devoid of regenerative capacity. Moreover, the brain's fragile neuron-based recording system holds dominion over uncountable behaviors essential for survival and reproduction. Thus it should not be surprising that viruses infecting the CNS can cause neurobehavioral sequela that range broadly in severity. At the far end of the spectrum are lytic flaviviruses, which can cause fatality rates in excess of 75% when they invade the CNS, but leave the host unfazed when their replication is confined to the periphery (Monath et al. 1996).

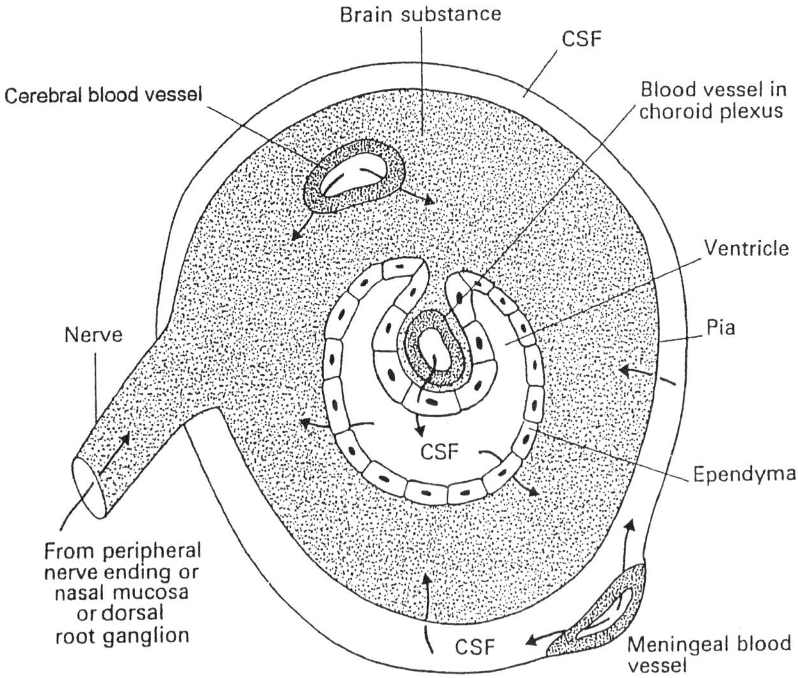

Figure 7. Routes of viral invasion of the central nervous system. CSF, cerebrospinal fluid. Reproduced with permission from: Scheld, W.M., Whitley, R.J., & Durack, D.T., ed. (1997) *Infections of the central nervous system, 2nd edition* (pp. 12). Philadelphia: Lippincott-Raven Publishers.

While its clear that viruses can have detrimental, even fatal, consequences if they breach the barriers protecting the central nervous system, a discussion limited to the acute effects of lytic viruses on vegetative functions is unlikely to bear directly on the current infectious hypothesis for psychiatric disorders. Fortunately, highly lytic viruses that are also neuroinvasive are relatively rare. But with regard to the viral hypothesis of schizophrenia, the last several decades have revealed new insights into the manifold means by which viruses can disrupt the integrity and function of neural systems.

2.1.5 Effect of Viruses on Nerve Cells

Broadly speaking there are two ways a viral infection can impact neural cells. First, the virus can act to directly perturb cell physiology. For example, some viruses deplete cellular metabolic pools or disturb ion gradients. Apoptosis is increasingly recognized as a mechanism by which several viruses cause neuronal death (Fazakerley & Allsopp 2001). This may serve an altruistic function for the host if it occurs prior to the virus completing it's replication cycle. On the other hand, if the virus can stave off apoptosis until mature virions are produced, then this discreet type of cell death may benefit the virus by allowing release of progeny while delaying local host inflammatory responses.

Inflammation commonly accompanies viral infections (Bilzer et al. 1996). This brings us to the second method by which viruses result in perturbation of nervous system tissues. Most viruses trigger an immune response, and commonly it is the immune effectors or molecules rather than the virus itself that are responsible for most of the tissue damage. The immune response may entail a direct attack on infected neural cells, and the inflammation produced by this response may go on to injure nearby cells that are mere bystanders to the main assault. More indirectly, the CNS can suffer the consequences of a viral infection that is localized in the periphery. For example, in parainfectious or postinfectious encephalitis the immune response is incited against a virus localized exclusively in the periphery, but in a manner that is not well understood, immune reactions are misdirected against cells residing within the CNS. This represents just one of a multitude of ways that the delicate dynamics of immune recognition and response to viral infections can go awry, and ultimately end up compromising the integrity and function of the brain.

The interplay between antiviral immunity and host tissue injury is central to many infectious theories of schizophrenia. Therefore the next section adumbrates those aspects of host immunity that are needed as a foundation for later discussion of immune mechanisms in the etiopathogenesis of schizophrenia and other psychiatric disorders.

2.2 IMMUNE RESPONSE TO VIRAL INFECTIONS

There have been hundreds of studies probing immune abnormalities in schizophrenia (Ganguli et al. 1993; Muller et al. 2000; Wright et al. 2001; Yolken et al. 1995). This should be considered in light of the fact that viral infections in humans almost always engage the immune system in some way. However, establishing a link between detected immune abnormalities in schizophrenia (or other psychiatric disorders of the unknown etiology) and viral infection is not a straightforward endeavor, and any clarity that may be achieved is dependent on understanding normal antiviral immune responses.

The body is absolutely dependent on efficient function of the immune system as illustrated starkly by the brief life expectancy (only months) of untreated infants born with severe combined immune deficiency disease. A reminder that the immune system operates continuously throughout the life span is provided by the fatal opportunistic infections and cancers that arise during AIDS. The immune system is charged with diligently surveying every tissue for foreign invaders and neoplastic cells. To help ensure a swift and decisive victory over a viral pathogen, the immune system launches a coordinated response involving mobile cells and chemical messengers that interact extensively. Specialized lymphatic tissues such as the lymph nodes and spleen serve as a staging ground for the immune operation.

Despite the magnificent ability of a healthy immune system to combat the interminable onslaught of viruses intent on gaining control of host cell functions, it would be erroneous to believe that immune responses have evolved to apply a perfectly balanced force that minimizes injury to host tissue while still clearing the virus. Indeed, many of the miseries we experience during a viral illness can be attributed to immune mediators and not to the virus. Thus, even a well functioning immune system readily causes tissue injury -- and in fact, some immune effectors are expressly designed to cause such damage (as discussed below). It is tempting to view autoimmune disease, whereby immune cells and soluble mediators are misguided toward self antigens, as being at the far end of a continuum just beyond the normal limits of an appropriately-vigorous immune response. But this is an oversimplification, and most autoimmune mechanisms are poorly understood. There is, nonetheless, a growing consensus that viruses can trigger chronic autoimmunity in susceptible humans, and the brain seems to be vulnerable to this process (Arnason 1998; Oldstone 1998). Whether an exposed individual develops such disease is presumably a function of pathogen variables as well as host immunogenetics and prior infectious disease history.

Autoimmune mechanisms have been frequently proposed to play a causal or contributing role in schizophrenia, major depression, and chronic fatigue syndrome. These mechanisms are considered further in Chapter 3. The immunological foundation for such a discussion of abnormal immune responses is closely tied to normal immune processes and is the subject of the following sections.

2.2.1 Components of the Immune System

The precise list of "immune system components" depends on how inclusive one chooses to be concerning the host elements participating in anti-microbial defense. Certainly, the immune system components can be extended beyond the specialized cells and organs to include external barriers (skin, mucous membranes), non-specific anti-microbial substances in body fluids (lysozyme in tears), and even the host's normal bacterial flora. Because these later "extended" components of the immune system are constitutive factors contributing to host defenses, they could be involved in individual vulnerabilities to viral infections, and hence could hold clues to potential viral mechanisms in psychiatric disorders. However, such factors have not been invoked as essential elements of the viral hypothesis, and for the purposes of this chapter, discussion will focus on the main cells and tissues responsible for anti-viral immunity.

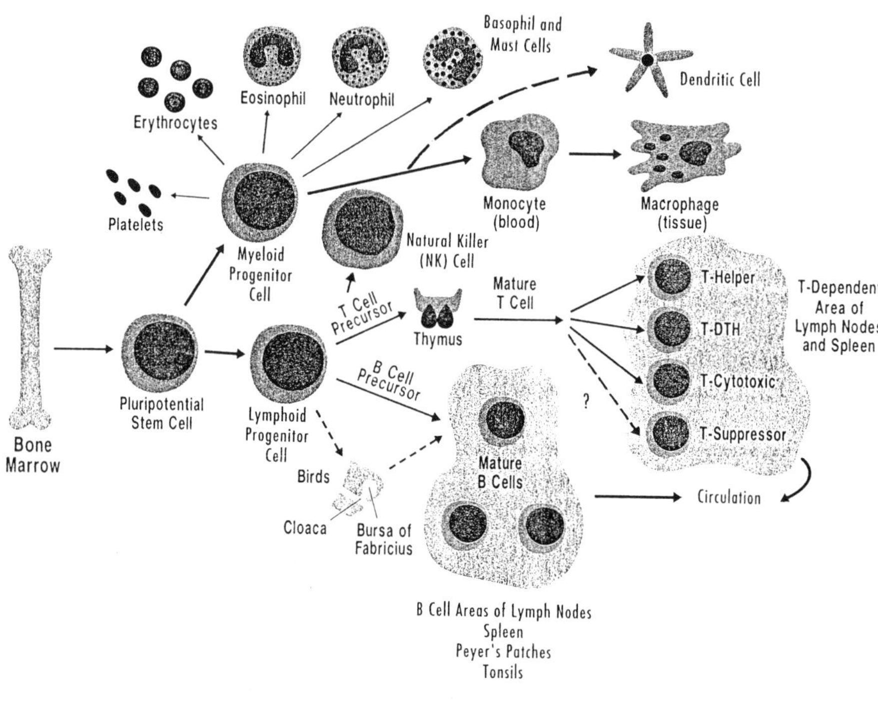

Figure 8. The developmental pathway of various cell types from a pluripotent bone marrow stem cell. Reproduced with permission from: Benjamini, E., Sunshine, G., & Leskowitz, S. (1996) *Immunology: a short course, 3rd edition* (pp. 189). New York: Wiley-Liss, Inc.

The chief organs of the immune system are bone marrow, spleen, thymus, lymph nodes, and lymphatic tissue of the gastrointestinal tract. As shown in figure 8, the major cells populating the adult immune system are derived from self-renewing stem cells in the bone marrow.

These pluripotent stem cells turn over constantly as they differentiated along "myloid" or "lymphoid" lineages under the influence of cytokines in the micro-environment of the bone marrow. Some of the cytokines responsible for orchestrating lineage decisions of immune cells are also critical participants in the differentiation of neural cells during embryogenesis (Mehler et al. 1995). This raises the intriguing possibility that anomalies in these cytokines (or other factors that regulate such cytokines) could underlie both the immune abnormalities and neurodevelopmental origins that have been proposed for schizophrenia.

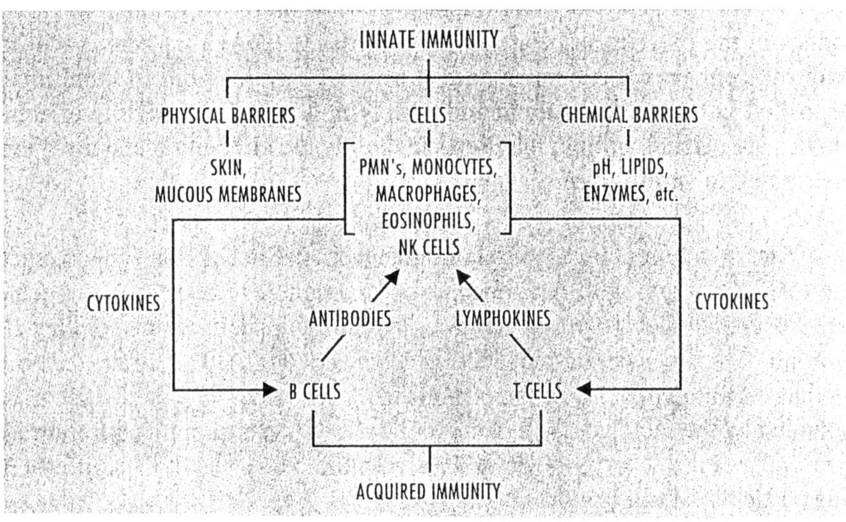

Figure 9. The interrlationship between the innate and acquired immunity. Reproduced with permission from: Benjamini, E., Sunshine, G., & Leskowitz, S. (1996) *Immunology: a short course, 3rd edition* (pp. 38). New York: Wiley-Liss, Inc.

2.2.2 Innate and Acquired Immunity

Functionally, immune responses are classified as either innate (natural) or acquired (specific) based on whether prior exposure to the pathogen is needed in order for the immune response to be fully armed and implemented

(Yewdell et al. 1997). Both subsystems play an indispensable role in defending against viral infections, and operationally, they interact and cooperate (Figure 9).

Immune responses to pathogens occur in waves, with initial recruitment of innate immune cells and molecules (NK cells, interferon) being followed by activation of acquired immunity (T-cells, antibody from B cells). Innate immunity is nonspecific in that it is not able to distinguish subtle molecular differences between pathogens, and it also lacks " memory " -- a poorly understood property that enables immune responses to be more abrupt and vigorous upon re-encountering a pathogen on a subsequent occasion. Even so, innate immunity represents an essential first line of defense against viral infections.

2.2.2.1 Innate Immunity

Specific cell types of the innate immune system include phagocytes (macrophages, neutrophils, and microglia) that engulf viruses, and natural killer cells that recognize infected cells (by unclear mechanisms) and destroy these newly corrupted virus-factories by releasing cytolytic molecules.

All cell types participating in innate immunity release cytokines (Biron et al. 1998). These soluble proteins serve to coordinate the early phase of the immune response and combat infection through direct antiviral activity (interferons), as well as by recruiting other immune cells to set-off the inflammatory cascade (interleukin-1, tumor necrosis factor alpha), and by priming later acquired responses (interleukin-12).

There are additional soluble mediators of innate immunity besides cytokines, including complement (discussed below) and other acute phase proteins. Over 30 acute phase proteins have been described, most of which are derived from the liver and are elaborated in response to pro-inflammatory cytokines and other soluble immune mediators. The finding of elevated plasma levels of acute phase proteins (e.g. C-reactive protein) is used clinically as a nonspecific indicator of inflammation. A number of reports have also found these proteins to be increased in the plasma of patients with schizophrenia (Maes et al. 1997; Rothermundt et al. 2001).

In general, however, innate immunity has received less attention in schizophrenia research than acquired immunity. Perhaps this is due to the intuitive notion that a disease as complex as schizophrenia-- affecting many of our highest intellectual functions-- would require an immune mechanism that is commensurately complex and highly evolved. Certainly innate immunity is evolutionarily less "advanced" than acquired immunity – which is the crowning achievement of mammalian host defenses.

2.2.2.2 Acquired Immunity

Microbial pathogens frequently mutate to avoid being recognized and destroyed by the immune system. Accordingly, the host must be highly adaptable in order to stay one step ahead of its microbial adversary. This is the duty of the T and B lymphocytes, which are responsible for acquired (or adaptive) immunity as they direct a tailor-made attack against foreign antigens. They accomplish this through antigen recognition: a complicated process whereby a B- or T- cell identifies short molecular bits of a foreign invader hence distinguishing it from the large background pool of self-molecules (Yewdell & Bennink 1997).

Specific immune responses can be conceptualized to occur in three phases: an induction phase, in which the presence of the virus is recognized; an activation phase, in which specific immune cells proliferate and are mobilized into action against the virus or its molecular constituents; and an effector phase, in which the infectious agents is neutralized or eradicated.

In order to recognize a viral antigen as non-self, both B- and T cells deploy specialized receptors on their surfaces that bind to identifying pieces of the viral protein or polysaccharide. The pieces of the molecular whole that are engaged by B or T cells are referred to as "epitopes." There are typically only a few such epitopes on a given viral protein (antigen). For B cell induction, the viral epitopes may exist free in solution (e.g. the exposed portion of a virion floating in plasma), or they may be on the surface of an infected cell. In any case, it is the idiosyncrasies of the immune system that determine precisely what is seen as an epitope (Yewdell & Bennink 1997).

The diversity of potential epitopes that can be recognized is enormous because--prior to encountering an antigen--each B-cell expresses a unique B-cell receptor molecule on its surface. These receptors are immunoglobulin structures that are similar to the antibody that the B cell progeny will ultimately produce. The extreme heterogeneity of the portion of these molecules responsible for binding epitopes is achieved by genetic recombination under control of recombinase enzymes. These enzymes were once thought to be expressed exclusively in B and T cells, but recent studies have found they are also expressed by neurons (Chun et al. 1991). On account of the diversity in B cell receptors there are millions of unique B-cells that the virus has to face, and when the B-cell receptor binds an epitope on the viral protein or polysaccharide, that cell becomes committed to producing large quantities of antibody that likewise recognize the same epitope.

Because proteins are folded in three-dimensional space, B cell epitopes need not consist of contiguous amino acid residues. It is important to distinguish between these conformational (or discontinuous) epitopes, and linear (or sequential) epitopes; the latter being formed entirely from adjacent amino acid residues (figure 10). This becomes relevant to neuropsychiatric disorders when considered in light of numerous studies evaluating patient's sera for antibodies that bind selectively to viruses or self-antigens (discussed in chapter 4). Thus, during laboratory procedures used to measure specific antibodies, the antigenic proteins are likely to be denatured (e.g. as in the standard western blot), and the recognition of conformational epitopes will not be detected.

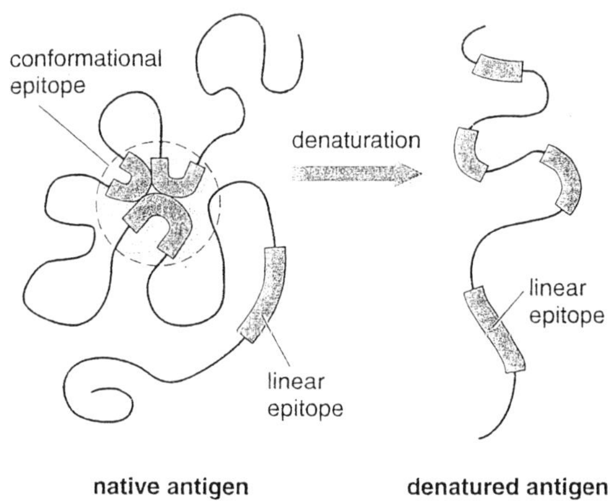

Figure 10. Conformational and linear epitopes in a polypeptide antigen. After denaturation, the conformational epitope can no longer be recognized by antibodies, but the linear epitope is unaffected. Reproduced with permission from: Parslow, T.G., Stites, D.P., Terr, A.I., & Imboden, J.B., ed. (2001) *Medical immunology* (pp. 75). New York: McGraw-Hill Companies, Inc.

For antigen recognition by T-cells there is an added level of complexity not inherent to B cells. The epitopes engaged by T-cells are antigen fragments that must be present in the association with a group of cell surface molecules called major histocompatability complex antigens (MHC) (Doherty 1997). There are two types of MHC, termed class I and class II. With the exception of neurons (discussed below) virtually all cells of the body express MHC class I (MHC I). The expression of MHC class II is limited primarily to cells that are intimately associated with immune responses. There is a variety of such MHC II –expressing cells. These share the ability to engulf, degrade, and process extracellular proteins, and then display these processed pieces in conjunction with MHC class II molecules. Such cells are called "antigen presenting cells", or "APCs."

In contrast to B cells, the induction of T cell responses depends on the ability and effectiveness of MHC molecules to bind and present short stretches of foreign proteins. T-cells are hence said to be "MHC restricted", because the antigen receptor on the surface of the T-cells (called the T cell receptor, or TCR) can only bind to antigen if it is associated with one or the other class of MHC molecule.

Two major types of T-cells can be distinguished based on their functional characteristics and whether they respond to antigen presented by class I or class II MHC. T-helper (CD4 positive) lymphocytes recognize antigen presented by class II, where as cytotoxic T (CD8 positive) lymphocytes respond only to antigen presented by class I. The role of the helper cell is to secrete cytokines that regulate activation and effector functions of other immune cells, including B cells. The contribution of CD8+ cytotoxic T lymphocytes is more direct: these cells bind to and lyse virus-infected cells consequently limiting the spread of the infection. Although the primary amino acid sequence of MHC I and MHC class II (MHC II) are not very similar, these molecules function in an analogous manner. Both are capable of presenting a diverse array of peptide fragments to T-cells. The genes conferring the capacity for the structural diversity needed to do this reside in a large gene complex that is located on chromosome 6. For historical reasons, this gene complex in humans is referred to as the HLA complex.

MHC II consists of two non-covalently associated polymorphic subunits (alpha and beta chains), whereas MHC I consists of a polymorphic alpha chain associated with a non polymorphic subunit (beta-2 microglobulin) that is encoded on a different chromosome. As illustrated in figure 11, the cellular pathways by which MHC I and MHC II pick-up their pathogen peptides for presentation to T cells is consistent with the respective roles of these cells in activating T-helper (CD4 positive) lymphocytes and cytotoxic T lymphocytes (CD 8 positive). MHC II engages T helper cells, and it presents antigen fragments from organisms that have been engulfed from the extra-cellular environment by antigen presenting cells. In contrast, MHC I binds to antigens produced inside the cell such as portions of incipient viral proteins. It presents these fragments to CD8 positive precursors of cytotoxic T-lymphocytes. The region of the MHC molecules that bind to the antigenic peptides are at the tips of each molecule, placing the MHC-peptide complex farthest away from the cell membrane to allow engagement by TCRs.

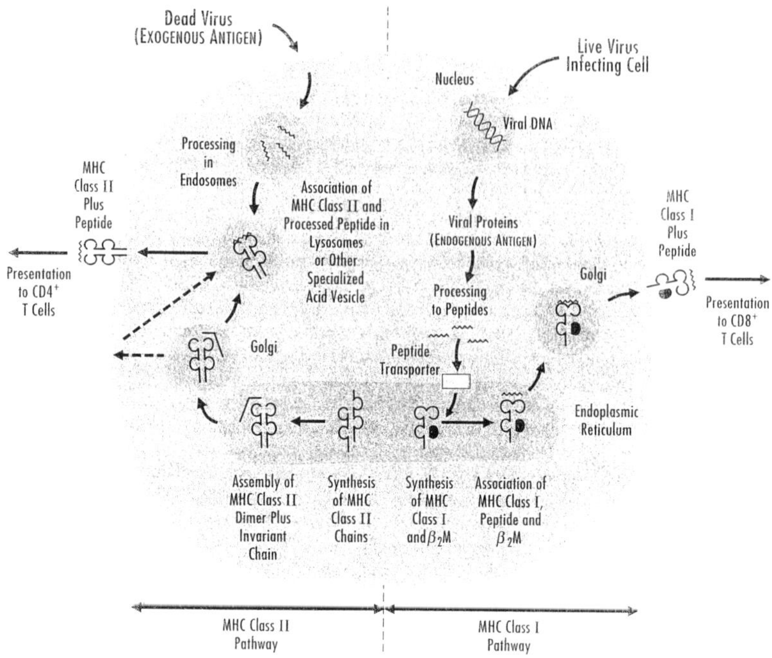

Figure 11. Differential processing of antigens in the MHC class II pathway (left) or MHC class I pathway (right). Dotted lines: "empty" class II molecules may move to the cell surface and cycle back to acid vesicles. Reproduced with permission from: Benjamini, E., Sunshine, G., & Leskowitz, S. (1996) *Immunology: a short course, 3rd edition* (pp. 189). New York: Wiley-Liss, Inc.

Similar to B-cells, the highly hyper-variable T cell receptor on T-cells allows for the recognition of a vast number of potential antigens. All the receptors on a particular T cell are identical and unique to that cell and its descendants. The descendants of a given T or B cell are called "clones"; these can be studied in the laboratory by isolating clonal cell lines, which maintain the antigen specificity of their predecessors.

MHC molecules are made by the most polymorphic genes in the human species. On a population level, this broadens the spectrum of viral peptides that can be presented to T-cells, and makes it difficult for pathogens to mutate and evade recognition. Accordingly, MHC and TCR molecules work in concert as a dual diversity system designed to circumvent the ability of pathogens to escape immune detection through mutations. Each person has three MHC I chains and three MHC II genes on each chromosome. The process by which these genes code for the various MHC molecules is depicted in figure 12. An individual who is homozygous would produce three MHC I, and three MHC II molecules-- each molecule having it's own preferential repertoire of peptides to present. However, because MHC genes are so polymorphic at the population level, it's very unlikely for someone to be homozygous. Moreover, the genes from both chromosomes in a pair are expressed co-dominantly on MHC-bearing cell. Consequently, individuals are likely to express six different MHC I, and six MHC II molecules, hence increasing the number of peptides that can be presented.

A given T cell will respond only to the appropriate combination of MHC and peptide. Accordingly, the inventory of MHC genes a person inherits determines which antigens will be presented, and ultimately strongly influences susceptibility to infectious diseases and autoimmune disorders. It also determines which peptides will be recognized as "non-self." Therefore, MHC matching (also called HLA tissue typing) is performed prior to transplantation to avoid tissue rejection.

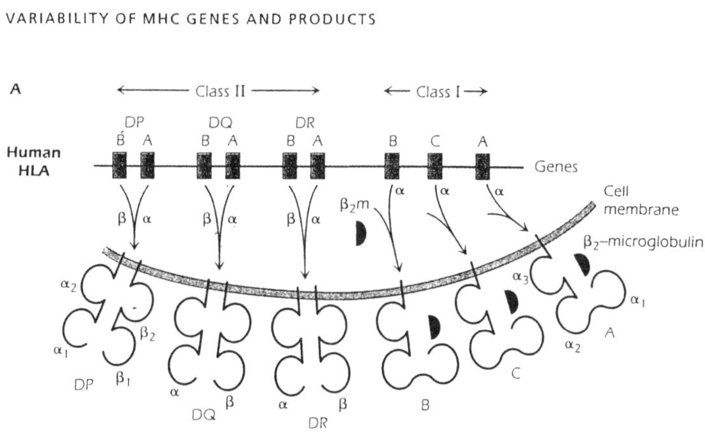

Figure 12. Simplified depiction of the human MHC, showing regions and genes coding for polymorphic MHC class I and II molecules. b2m = b2 – microglobulin, encoded outside the MHC. Reproduced with permission from: Benjamini, E., Coico, R. & Sunshine, G. (2000) *Immunology: a short course, 4th edition* (pp. 149). New York: Wiley-Liss, Inc.

Predictably, the realization that MHC molecules are perfectly poised to mediate interactions between genetic variables and environmental factors (namely infections) has fuelled association studies between MHC types and schizophrenia. Indeed, since the first such report in 1974 there have been over 60 studies examining the relationship between MHC alleles and schizophrenia prevalence or disease subtypes (Cazzullo et al. 1974; Wright et al. 2001). These studies are considered in more detail later on. The current chapter will outline the role of MHC molecules in immune responses to viruses.

Binding of a B or T cell antigen receptor to its cognate epitope is the first stage in the activation phase of specific immunity. Accessory signals, generated by cytokines derived from antigen presenting cells and T helper cells, cooperate to ramp-up the activation process. A cascade of cellular and molecular interactions becomes entrained as the host marshals the immune response to fend off infection. During this process T helper cells play a central role. Recent evidence indicates that there are two classes of T helper cells that can be distinguished based on their cytokine profiles after stimulation. The TH1 subtype secretes IL-2 and interferon-gamma, which activate cell mediated immune responses and induce B cells to make a class of antibody that utilizes complement in effector functions. TH2 cells secrete a different set of cytokines that encourage the production of antibodies that neutralize antigen upon binding. Recent studies have begun to investigate the idea that schizophrenia is associated with an imbalance between TH1 and TH2 cytokines (Muller et al. 1999). At present, however, these data are far from conclusive.

After T and B lymphocytes are activated, clonal expansion (proliferation) leads to a large pool of cells with identical high affinity and specificity for the pertinent antigens. Some of the progeny differentiate into mature effector cells such as antibody secreting plasma B cells or fully activated cytotoxic T lymphocytes. Conversely, other descendants of activated B- or T cells become memory cells that are primed for a more rapid and accentuated response upon re-exposure of the host to the same. The result of such memory cells is that exposure to the same viral antigen on a second occasion results in a more rapid and amplified response (as shown for B-cells in Figure 13). These memory cells last in the body for years, or even the entire life span, and they are responsible for generating a secondary immune response as is seen following a vaccine booster.

During the effector phase of a successful immune response the pathogen is neutralized or eliminated. The predominant mediators during this phase are antibodies generated from plasma cells, and CTLs. Different classes of antibodies (i.e. IgG, IgA, IgM) sub serve different functions during a viral infection (table 2). One of the main functions of antibodies is to bind to and coat virions consequently inactivating infectivity. IgG antibodies create a link with natural immunity because viruses coated with IgG become targets for lysis by complement proteins and are more easily phagocytized. Antibodies also bind to the surface of infected cells, marking them for destruction by NK cells. Even so, the selective destruction of infected cells is the main job of CTLs, which grasp infected cells and destroy them by releasing cytolytic molecules. This process is very efficient at limiting the spread of infection. A multicellular host will normally accept the price of destroying some individual cellular members of its legion in order to keep the virus in check. But what about cells that cannot be replaced? And how does the host cope with pathogens that have evolved to be minimally disruptive "guests." These points are considered in the next section.

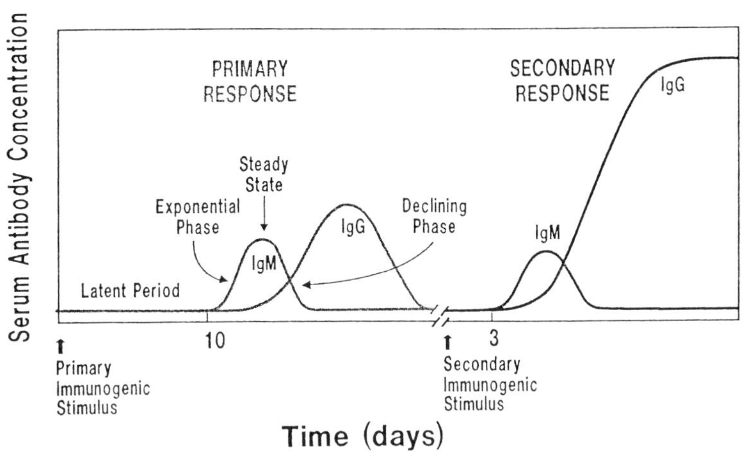

Figure 13. The kinetics of an antibody response. Reproduced with permission from: Benjamini, E., Sunshine, G., & Leskowitz, S. (1996) *Immunology: a short course, 3rd edition* (pp. 88). New York: Wiley-Liss, Inc.

Table 2 Properties of immunoglobulins important in viral infections.

	IgG	IgA	IgM
Major function	Virus clearance from extracellular spaces	Virus inactivation on mucous membrane	Virus clearance from blood
% Total immunoglobulin	80	13	6
Concentration in serum (mg/mL)	8–16	1.4–4	0.5–2
Cerebrospinal fluid–serum ratio	1/200–1/400	1/200–1/400	1/1,000
Half-life (days)	25	5	2–5
Transmission to offspring	Transplacental	Breast milk	0
Structure			
Heavy chain	$\gamma1$, $\gamma2$, $\gamma3$, $\gamma4$	$\alpha1$, $\alpha2$	μ
Light chain	κ or λ	κ or λ	κ or λ
Formula	$2\gamma + 2\kappa$ or 2λ	$2\ (2\alpha + 2\kappa$ or $\lambda) +$ J chain + secretory piece	$5\ (2\mu + 2\kappa$ or $\lambda) +$ J chain

Ig, immunoglobulin.

Reproduced with permission from: Johnson, R., ed. (1998) *Viral Infections of the Nervous System* (pp. 70). Philadelphia:, Lippincott Williams & Wilkins.

2.3 PERSISTENT VIRAL INFECTIONS.

In the classic view of viral diseases there only two possible outcomes for the host: either it rids itself of the infection or dies trying. However, this view has been reformulated according to the findings of modern virology, which takes into account the complex tactical battle between a virus and its host. This battle is waged on both an individual and evolutionary scale, with the immune system and brain playing a crucial role (Buchmeier et al. 2001).

It is now known that many viruses can evade immune recognition, at least to an extent, and consequently established a tenuous peace with their hosts (Ahmed et al. 1997; de la Torre et al. 1996). Hence, each of us unknowingly harbors some of these tiny infectious "guests." The viruses may reside in our immune cells, nervous tissue, or gut. This realization has opened new vistas in viral pathogenesis research, and has created a more accepting environment for novel pathophysiological paradigms linking infections with neurological and psychiatric diseases of unknown etiology.

While the majority of viruses infecting humans are cleared by the immune system, some viruses can linger or persist for an extended period of time-- months, years, or even the entire normal human life span. Viruses have developed various strategies to achieve persistence (Table 3).

As can be seen in figure 14 there are essentially four patterns of infection that have been identified thus far 1) *Acute self-limiting infections* are cleared by the immune system and therefore the virus remains subsequently undetectable. The morbidity that arises from these infections is assumed to be immediately evident, but this is not necessarily the case, especially in instances when the infection occurs in the context of development, or if it triggers an autoimmune disease. 2) The second pattern characterizes *chronic infections* (e.g. hepatitis C); the virus continues to be shed by infected tissues, albeit usually at lower levels than during the preceding acute infection. Chronic infections can be clinically silent, or the signs and symptoms of disease can arise immediately after infection, or following a considerable delay. 3) Latent infections (e.g. herpes simplex) are typified by periods of the viral quiescence, during which infectious virus is not detectable; these are punctuated by periods of reactivation and symptomatic recrudescence, during which the virus can usually be recovered. Thus, latent infections are caused by viruses that can cycle between productive and nonproductive modes. 4) *Slow infections* produce disease after a long incubation period of years or even decades. Slow infections are normally associated with transmissible spongiform encephaolopathies (Kuru, Creutzfeldt-Jakob disease), which aren't technically caused by viruses but rather are elicited by a proteinacious infectious particle termed a "prion"

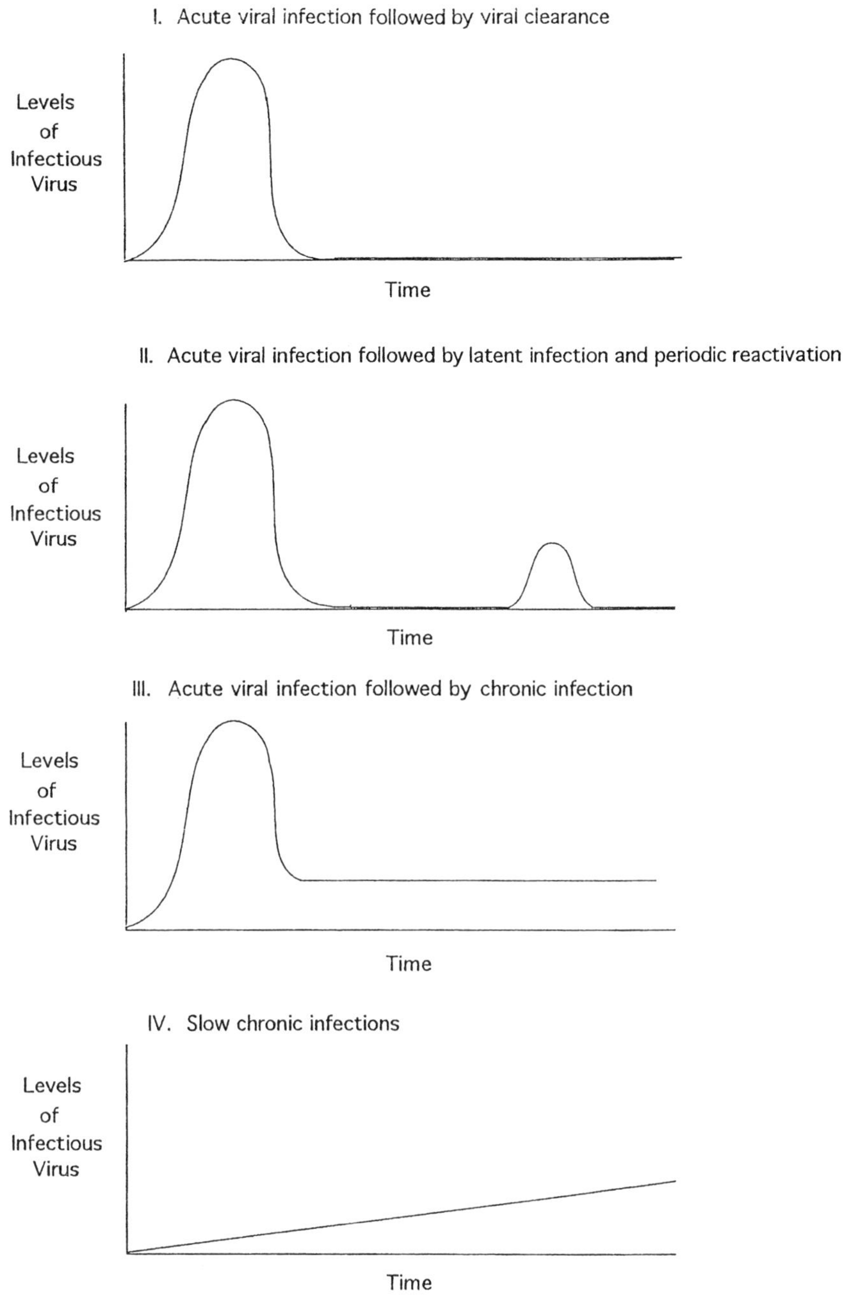

Figure 14. General patterns of viral infection in the host organism. Reproduced with permission from: Ahmed, R., Morrison, L.A., and Knipe, D.M. In: Nathanson, N., ed. (1997) *Viral pathogenesis* (pp. 221). Philadelphia: Lippincott-Raven Publishers.

Table 3 Essential requirements for viral persistence.

I. Avoidance of immunological surveillance
  A. Removal of recognition molecules from surface of infected cells
    1. Alteration of viral protein expression, antiviral antibody response inducing capping and modulation of viral gene expression
    2. Alteration of MHC expression
      a. Direct viral effect on MHC expression
      b. Modulation of MHC expression by virally-induced cytokine production
      c. Negligible expression in cells that do not have potential for regeneration (i.e., neurons)
  B. Infection of thymus prior to or during immune system development (central tolerance or hyporesponsiveness) establishing persistent infection
  C. Infection of peripheral immune system (peripheral tolerance, energy, immunological ignorance)
    1. Abrogation of lymphocytic, macrophage, and antigen-presenting function
      a. generalized immunosuppression
      b. Selective immunosuppression
      c. Lymphocyte/monocyte/macrophage/cytokine-induced alteration of gene translation
  D. Generation of antibody or cytotoxic T-lymphocyte escape variants

II. Noncytolytic Replication
  A. Generation of viral variants
    1. Defective interfering (DI) viral particles
    2. Single or limited amino acid mutation
  B. Diminished expression of viral gene product

Reproduced with permission from: De Le Torre, J. & Oldstone, M. Viral Persisitence. Advances in Viral Research 1996, 46: 313.

The central nervous system is particularly conducive to harboring persistent viral infections because its structural features and unique immunological characteristics permit viruses to be camouflaged from immune surveillance (Ahmed et al. 1997; de la Torre & Oldstone 1996). The brain has historical a been considered "immune-privileged" because it lacks conventional lymphatics, and the blood brain barrier serves to limit the entry of antibodies, cytokines, and other immune molecules present in the peripheral circulation. Similarly, immune cells enter the brain and exert their effector functions under a more restrictive set of rules than apply to the periphery (Dorries 2001). Moreover, neurons do not express MHC I, presumably because a CTL-mediated destruction of these (generally) irreplaceable cells would be counterproductive to host vitality. Thus neuronal infection by a noncytolytic virus presents the host with a gambit: the invader is reluctantly welcomed because its presence is preferable to the immune mediated destruction of essential and irreplaceable neurons.

In this regard, animal studies have firmly established that noncytolytic viruses can persistently infect neurons without producing pathological or cytoarchitectural alterations, and without inciting a local immune response (de la Torre & Oldstone 1996). However studies that began in 1970's have now established that there are often additional trade-offs for silently harboring a persistent viral infection. These studies (using predominantly animal models and in vitro analysis) have found that despite the absence of structural alterations or immune infiltration, persistent noncytolytic viruses can perturb differentiated or "luxury" functions of neurons, including neurotransmitter production. Indeed, a large body of evidence has accumulated demonstrating the capacity of chronic or persistent viral infections to alter differentiated or specialized cell functions without compromising cell survival (table 4).

These findings may be relevant to the pathogenesis of a number of diseases, including cancer, and they spurred somewhat of a revolution in the paradigms relating viral infections to psychiatric or neurological diseases. For example, because a persistent viral infection could occur in the brain without leaving any of the hallmarks of an infectious process, the relevant neuropathological findings in schizophrenia-- the absence of immune infiltration, lack of vial inclusion bodies, and minimal gliosis-- can no longer be considered definitive evidence against the viral hypothesis. Furthermore, the ability of noncytolytic viruses to disrupt neurotransmitter production and dampen hormones involved in neurodevelopment seems to fit well with a disease such as schizophrenia, which is not characterized by the wholesale loss of cells, but rather appears to involve subtle disruption of neuronal function or brain development. This being said, there is no conclusive evidence in humans that disruption of differentiated cell function plays a role in any neuropsychiatric disease.

Table 4 Viral infections interfering with differentiated cell function leading to disordered homeostasis and disease.

Syndrome phenotype	Cell tissue target	Virus	Reference
A. Tissue culture			
Decreased acetylcholinesterase and choline acetyltransferase	Neuroblastoma	Lymphocytic choriomeningitis virus (LCMV)	Oldstone *et al.* (1976)
Altered synthesis of melanin, mucosin, and sulfated proteoglycan	Melanoblastoma, myoblastoma, chondroblasts	RSV/chicken cells	Holtzer *et al.* (1975), Roby *et al.* (1976)
Alered lipid metabolism	BGM cells	Measles virus (MV)	Wild *et al.* (1981), Anderson *et al.* (1983)
Changes in phospholipid methylation	Glioma cells	Canine distemper virus (CDV), MV	Balbach and Koschel (1979), Koschel and Muntel (1980), Muntel and Koschel (1982)
Altered NK activity and immunoglobulin synthesis	NK and B cells	Cytomegalovirus (CMV), MV	Casali *et al.* (1984)
Altered MHC expression	MHC expressing cells susceptible to adenovirus	Adenovirus	Wold and Gooding (1991)
Diminished CTL activity	—	CMV	Krajcsi *et al.* (1992)
Altered hormone secretion	Somatotropes	Moloney leukemia virus (M-MuLV)	Schrier and Oldstone 1986)
Changes in cell morphology and cell gene expression	Fibroblasts	Foot-and-mouth disease virus (FMDV)	de la Torre *et al.* (1988)
B. *In vivo*			
Growth retardation: reduced growth hormone (GH) mRNA and protein levels	Growth hormone-producing cells in anterior pituitary	LCMV/mouse	Oldstone *et al* (1982, 1984)
Obesity: reduced levels of norephinephrine and dopamine	Hypothalamus involvement	CDV/mouse	Lyons *et al.* (1982), Bernard *et al.* (1983)
Deformed whiskers	Unknown	MCF-recombinant virus/mouse	Rowe (1983)
Diabetes: Hyperglycemia, abnormal glucose tolerance test	β-cells in islets of Langerhans	Venezuelan encephalitis virus (VEV)/hamster, LCMV/mouse	Rayfield *et al.* (1981), Oldstone *et al.* (1984)
Hypothyroidism: reduced levels of T3 and T4, decreased levels of thyroglobulin mRNA	Thyroid follicular cells	LCMV/mouse	Klavinskis and Oldstone (1986)
Neurochemical and neurobehavioral alterations	Neurons, astrocytes, other CNS cell types	Semliki forest virus (SFV)/mouse, LCMV/mouse, Rabies virus (RV)/rat/mouse, Borna disease virus (BVD)/rat	Barrett *et al.* (1986), Hotchin and Seegal (1977), Lipkin and Oldstone, Gold *et al.* (1994), Tsiang (1992), Dittrich *et al.* (1989), Bautista *et al.* (1994)

Reproduced with permission from: De Le Torre, J. & Oldstone, M. Viral Persisitence. Advances in Viral Research 1996, 46.

There is considerable work to be done in order to ascertain the potential contribution of persistent and chronic viral infections in neuropsychiatric disorders. This is a difficult task, and even for neurological diseases with a known viral cause there is only a rudimentary understanding of the temporal and pathophysiological association between the persistent infection and the disease states that are engendered. Nevertheless, it is clear that some viruses can lie dormant in an occult form only to produce clinical manifestations much later. For instance, rubella can cause an insidious encephalitis that is manifested clinically as intellectual decline accompanied by other behavioral anomalies. The syndrome is rare, but patients are typically asymptomatic for about a decade following the initial infection, and while death usually occurs within several years of disease onset, some patients undergo a remission of the neurobehavioral symptoms. Measles can also produce prominent psychiatric features. In one form of the disease (subacute sclerosing panencephalitis) the average latency period between infection and clinical manifestations is seven years.

In summary, there is no doubt that viruses can cause neurobehavioral alterations after considerable latency in humans. Viruses also clearly have the capacity to disrupt subtle functions of neural cells, and accomplish this while avoiding immune surveillance. Subsequent chapters will provide additional coverage of these areas. But before leaving the subject of general virology and immunology, it is important to emphasize that there has been an explosion of knowledge in these fields in the last thirty years, resulting in countless new avenues for exploring the viral hypothesis of schizophrenia. For the most part, these have remained un-traveled.

REFERENCES

Ahmed R, Morrison LA, Knipe DM.Viral Persistance. In: N Nathanson; R Ahmed, editors Viral pathogenesis. Philadelphia: Lippincott-Raven; 1997; p. 181-205.

Arnason BGW.Autoimmune diseases of the central and peripheral nervous systems. In: NR Rose; IR Mackay, editors The Autoimmune Diseases. San Diego: Academic Press; 1998; p. 571-602.

Bilzer T, Stitz L. Immunopathogenesis of virus diseases affecting the central nervous system. Critical Reviews in Immunology 1996; 16:145-222.

Biron CA, Cousens LP, Ruzek MC et al. Early cytokine responses to viral infections and their roles in shaping endogenous cellular immunity. Advances in Experimental Medicine & Biology 1998; 452:143-9.

Brooks GF, Butel JS, Morse SA et al. Jawetz, Melnick, & Adelberg's medical microbiology. 22nd ed. New York: Lange Medical Books/McGraw-Hill Medical Pub. Division; 2001. , 694 p..

Buchmeier MJ, Cambell IL. Neurovirology: Viruses and the Brain. K Maramorosch; FA Murphy; AJ Shatkin, editors. San Diego: Academic Press; 2001. 645 p. (K Maramorosch; FA Murphy; AJ Shatkin editors. Advances in Virus Research).

Cazzullo CL, Smeraldi E, Penati G. The leucocyte antigenic system HL-A as a possible genetic marker of schizophrenia. British Journal of Psychiatry 1974; 125:25-7.

Chun JJ, Schatz DG, Oettinger MA et al. The recombination activating gene-1 (RAG-1) transcript is present in the murine central nervous system. Cell 1991; 64:189-200.

de la Torre JC, Oldstone MBA. Anatomy of viral persistence: mechanisms of persistence and associated disease. Advances in Virus Research 1996; 46:311-43.

Dermody TS, Tyler KL.Viral Diseases. In: GL Mandell; RG Douglas; JE Bennett et al., editors Mandell, Douglas, and Bennett's principles and practice of infectious diseases. 5th edn. Philadelphia: Churchill Livingstone; 2000; p. 1536-2986.

Doherty PC. The Nobel Lectures in Immunology. The Nobel Prize for Physiology or Medicine, 1996. Cell mediated immunity in virus infections. Scandinavian Journal of Immunology 1997; 46:527-40.

Dorries R. The role of T-cell-mediated mechanisms in virus infections of the nervous system. Current Topics in Microbiology & Immunology 2001; 253:219-45.

Fazakerley JK, Allsopp TE. Programmed cell death in virus infections of the nervous system. Current Topics in Microbiology & Immunology 2001; 253:95-119.

Ganguli R, Brar JS, Chengappa KNR et al. Autoimmunity in schizophrenia: a review of recent findings. Ann. Med. 1993; 25:489-96.

Johnson RT. Viral infections of the nervous system. 2nd ed. Philadelphia: Lippincott-Raven; 1998. xiv, 527 p.

Khabbaz RF. Emerging Viral Infections. Advances in pediatric infectious diseases 1999; 14:1-27.

Maes M, Delange J, Ranjan R et al. Acute phase proteins in schizophrenia, mania and major depression: modulation by psychotropic drugs. Psychiatry Research 1997; 66:1-11.

Mehler MF, Kessler JA. Cytokines and neuronal differentiation. Critical Reviews in Neurobiology 1995; 9:419-46.

Monath TP, Heinz FX.Flaviviruses. In: BN Fields; DM Knipe; PM Howley, editors Fields Virology. Vol. 1. Philadelphia: Lippincott-Raven; 1996; p. 961-1034.

Muller N, Riedel M, Ackenheil M et al. The role of immune function in schizophrenia: an overview. European Archives of Psychiatry & Clinical Neuroscience 1999; 249 Suppl 4:62-8.

Muller N, Riedel M, Gruber R et al. The immune system and schizophrenia. An integrative view. Annals of the New York Academy of Sciences 2000; 917:456-67.

Oldstone MB. Molecular mimicry and immune-mediated diseases. FASEB Journal 1998; 12:1255-65.

Rothermundt M, Arolt V, Bayer TA. Review of immunological and immunopathological findings in schizophrenia. Brain, Behavior, & Immunity 2001; 15:319-39.

Samson M, Libert F, Doranz BJ et al. Resistance to HIV-1 infection in caucasian individuals bearing mutant alleles of the CCR-5 chemokine receptor gene. [see comments.]. Nature 1996; 382:722-5.

Tyler KL, Fields BN.Pathogenesis of viral infections. In: BN Fields; DM Knipe; PM Howley, editors Fields Virology. Philadelphia: Lippincott-Raven; 1996; p. 173-218.

Van Regenmortel MHV, Fauquet CM, Bishop DHL et al. Virus Taxonomy. Seventh report of the International Committee on Taxonomy of Viruses. San Diego: Academic Press; 2000.

Wright P, Nimgaonkar VL, Donaldson PT et al. Schizophrenia and HLA: a review. Schizophrenia Research 2001; 47:1-12.

Yewdell JW, Bennink JR.Immune Responses to Viruses. In: DD Richman; RJ Whitley; FG Hayden, editors Clinical virology. New York: Churchill Livingstone; 1997; p. 271-305.

Yolken RH, Torrey EF. Viruses, schizophrenia, and bipolar disorder. Clinical Microbiology Reviews 1995; 8:131-45.

Chapter 3

ENCEPHALITIS AND SCHIZOPHRENIA

3.1 VIRUSES AND SCHIZOPHRENIA: A PLURALITY OF THEORIES

The viral hypothesis for schizophrenia does not consist of a single line of reasoning, but rather is an accretion of assorted "clusters" of data that can be connected in various ways to form the often ill-defined sub-theories of the hypothesis. These sub-theory themes range from autoimmunity to the virology of retroviruses. and range considerably in their cogency, clarity, and relevance to schizophrenia. This chapter begins to scrutinize some of these or clusters of evidence and the theories that rely on them. Specifically, it will cover: i) the established ability of viruses to cause psychosis; and ii) the burgeoning and controversial literature on Borna disease virus.

The last section of this chapter, will provide a brief overview of the role of infections in other psychiatric disorders. The relevance of this literature to schizophrenia is twofold. Firstly, it adds plausibility to the idea that microbial organisms can interact with the brain and cause complex behavioral syndromes. Secondly, there may be co-morbidities or shared genetic liabilities with these diseases and schizophrenia, and thus by elucidating the roots of one disease we might may gain clues to the pathogenesis of the other.

Broadly speaking, if a virus is involved in schizophrenia, it could either act during development to initiate brain abnormalities that eventually give rise to psychotic symptoms, or the virus could act more proximally to the psychotic episode, as is seen in acute encephalitis. Because theories proposing a neurodevelopmental origin for schizophrenia have been gaining momentum, and seem to fit well with a viral insult, Chapter 5 explores viral mechanisms during development. The current chapter emphasizes viral and immune interactions with the mature brain. However, the clusters of the evidence supporting early versus late influences on the brain cannot be so readily sorted into conceptual bins of developmental versus proximal-causes. For example, serological evidence of increased exposure to herpes viruses in schizophrenia patients could be construed as a long-lasting vestige of a neonatal infection, or conversely could be interpreted as a rise in titer due to infection that occurred near the time of mental illness onset. Along similar lines, infection early in life may set the stage for an abnormal immune response that is mounted against a related viral infection occurring later in life. This cross-reacting second immune response could, in theory, be the proximal cause of the neurobehavioral abnormalities seen in schizophrenia. In this case, schizophrenia would only arise in individuals who were exposed to both viruses with the appropriate timing.

3.2 PSYCHOSIS IN VIRAL ENCEPHALITIS

Psychosis due to acute viral encephalitis is a well-established disorder that is classified according to the Diagnostic and Statistical Manual of Mental Disorders as psychotic disorder due to general medical condition, or delirium due to general medical condition (depending on accompanying neurobehavioral symptoms) (Lishman 1987; Little et al. 1998). The attribution of psychosis to infection is most straightforward when psychotic symptoms occur in close temporal proximity to the signs, symptoms and diagnostic indices of viral encephalitis (e.g. fever, headache, CSF pleocytosis, rising viral titer, EEG abnormalities, and culture-positive biopsy). However, it is worth noting that the diagnosis of viral encephalitis itself may often be made on a presumptive basis, and often the responsible microbe is not identified. For example, in a study of 432 patients who underwent brain biopsy for presumptive herpes simplex encephalitis, only 45% of the cases could be confirmed by isolation of the virus (Whitley 1997). Of the remainder, 7% of the cases were due to non-infectious causes and 6% were caused by bacterial or fungal-type infections. Nevertheless, viruses were clearly the most important causes of the encephalitis in the study, and 9% of cases were found to be due to various other viruses including St. Louis encephalitis, mumps, and influenza.

There is no doubt that viral infections of the central nervous system can cause psychosis. These symptoms may arise contemporaneously with focal neurological signs, but more complex patterns have also been well described. Torrey reviewed 22 cases in which patients were originally diagnosed with a psychotic disorder and later found to have evidence of a viral encephalitis (Torrey 1986). The presumptive agents responsible included herpes simplex virus, measles, Epstein Barr virus, cytomegalovirus, influenza, coxsackie, and Eastern equine encephalitis. Delusions and auditory hallucinations were present in most of the cases, but negative symptoms were relatively less common. The delusions were often bizarre; one man believed he was the leader of a Kingdom of insects, and another identified himself as Clark Gable from the planet Mars. Three of the 22 patients claimed to be dead. The lag between the development of psychotic symptoms and the neurological manifestations of encephalitis varied from days to several weeks. The ultimate clinical course ranged from death (5 patients) to complete recovery (8 patients); about one fourth of the surviving patients had residual psychiatric symptoms.

Psychosis can also occur as a sequela to a viral infection that actually never enters the brain. In this disease, which is been referred to by various names such as 'post infectious encephalomyelitis" and "acute disseminated encephalomylelitis", the peripheral infection triggers an inflammatory process in the central nervous system. This disease can also be triggered by vaccines. Myelin is extensively involved and focal neurological signs are usually evident. In rare instances however, this immune mediated disease can present as an acute psychosis.

The next section describes several specific viruses known to be associated with psychosis. This is by no means an exhaustive list of infections reported in the literature to present at some point during their course with signs or symptoms of psychosis.

3.2.1 Herpes Encephalitis

There are two types of the herpes simplex virus, type 1 and type 2 (Nahmias et al. 1973). These pathogens are among nine members of the Herpesvirdae family that are known to cause human disease. Other members included cytomegalovirus, Epstein-Barr virus, and varicella-zoster. Herpes encephalitis is a rare but serious complication of infection with HSV-1 or HSV-2. Any age group can be affected, but most congenital cases are caused by HSV-2 while the majority of other cases are caused by HSV-1. Although the annual incidence of herpes simplex encephalitis is only about 1 in 300,000, more than half of untreated cases are fatal, and most of the survivors have permanent neuropsychiatric sequela.

HSV-1 is a very common virus. The majority of people (80-90% in the United states) become infected with HSV-1 at some point in their lifetime (Whitley 1997). The virus initially infects the orolabial region, and then establishes latency in the trigeminal ganglion, and undergoes periodic reactivation causing "cold sores." Most adult cases of herpes encephalitis arise from these latent infections but the route by which the virus gains entry to the CNS parenchyma has not yet been deciphered. Typically, the onset of herpes simplex encephalitis occurs over several days. Patients usually have a headache and fever as well as behavioral abnormalities. Indeed, alterations in personality are evident in 85% of cases at initial clinical presentation. Thus, this abnormality of higher cognitive function is found more often in the early stage of the disease than dysphasia (76%), ataxia (40%), or seizures

(38%). Behavioral disturbances during the acute disease phase include confusion, as well as hallucinations that may be visual, olfactory, or gustatory (Lishman 1987; Wilson 1976). The characteristics of the symptoms are broadly consistent with the virus's tropism for orbitotemporal brain regions.

Because delusions, paranoia, or catatonia may dominate the early clinical picture of herpes encephalitis, some patients may be prematurely diagnosed with a schizophrenia spectrum disorder. Psychiatric symptoms often continue to evolve during the recovery phase of the disease. Besides persistence of memory deficits, recovering patients may develop aggressiveness, emotional lability, and other aberrant behaviors including hyperphagia. Long-term neuropsychiatric sequela are common even in patients who are successfully treated with anti-viral medication during the acute phase of the illness. McGrath et al. assessed persistent psychiatric abnormalities in 29 survivors of herpes simplex encephalitis (McGrath et al. 1997). Memory impairment was the most common disorder (69% of cases), but 45% of the cases had significant psychiatric morbidity including severe anxiety, depression, emotional liability, and obsessive-compulsive behavior.

Generally, herpes simplex encephalitis is considered a severe acute neurological disorder that leads to residual neuropsychiatric impairment. However, because HSV-1 shows a proclivity for nervous tissue and undergoes cycles of latency and reactivation, some authors have entertained the idea that HSV could be a cause of idiopathic psychosis, including schizophrenia. Such a mechanism rests substantially on the assumption that herpes simplex is capable of entering the brain and running a non-fulminant disease course. There is little convincing evidence for this, but still the concept has led to numerous attempts to isolate HSV from the brains of psychiatric patients, and has spurred a host of studies endeavoring to uncover serological evidence for such infections. These data are considered in subsequent sections.

3.2.2 Encephalitis Lethargica

Encephalitis lethargica was first reported in 1917 by von Economo who noted an unusual constellation of neurological and psychiatric symptoms in patients afflicted with an influenza-like illness during an epidemic in Austria (Lishman 1987; McKenna 1994). In the ensuing 10 years, the epidemic spread throughout Europe, affecting about 5 million people. Because of the severity of the neuropsychiatric sequela and the large number of people infected worldwide, the clinical characteristics of encephalitis lethargica were studied extensively. However, the science of virology was immature during this period, and the "filterable agent" (virus) that caused encephalitis lethargica was never definitively identified. Still, the epidemic was of great historical importance for its role in confirming that "functional" psychiatric disorders can have organic causes.

The geographically localized outbreaks of encephalitis lethargica during the 1920's tended to show a seasonal cyclicity, with peak incidence in early winter. The virus was said to target individuals in early adulthood, but no age group was entirely spared. The incidence peaked in the mid 1920's and faded away by the early 1930s. Over one-third of patients with well-defined encephalitis lethargica died from the illness or its complications. Of the survivors, approximately two-thirds were left with residual neuropsychiatric symptoms that were often severe and disabling.

The clinical course of encephalitis lethargica has been described in numerous journal reports as well as a book by Oliver Sacks. By definition, the disease entails neurological or psychiatric abnormalities. Typically, the central nervous system manifestations of the disease were ushered in by an influenza-like prodrome consisting of fever, pharyngitis, headache, and malaise. The neurological and behavioral abnormalities that followed varied somewhat between individual patients; and prominence of certain symptoms and signs could differ between the various geographic and temporally-spaced epidemics. Three forms of the disease were defined, but patients could manifest signs of more than one form, either simultaneously or sequentially (Lishman 1987). The somnolent-opthalmoplegic form was most common. Its main features were hypersomnolence and cranial nerve paresis. Less common was the hyperkinetic form, which presented with restlessness, insomnia, and purposeless movements. This form mimicked catatonic excitement. The parkinsonian form of the disease was the least common of the three. Patients with this form had a paucity of movement that could be observed as a waxy flexibility or persistence of postures, hence mimicking a catatonic stupor.

Occasionally, psychosis would either precede or coincide with the acute phase of encephalitis lethargica, but most often it became clinically apparent years after the onset of the viral illness. The available data from studies that examined psychiatric illnesses in encephalitis lethargica indicate that 15 to 30% of patients developed a schizophrenia-like psychosis (e.g. delusions, paranoia, or auditory hallucinations). However the exact number of psychotic phenomena occurring in unselected cases was never precisely determined. There will likely always be some mystery shrouding encephalitis lethargica because the responsible agent was never isolated, and the environmental conditions that gave rise to the epidemics of the 1920's were never determined. Additional neuropathological examination of archival nervous tissue from patients who died from the disease may hold some clues, and there is certainly no guarantee that our present day society will be spared another epidemic caused by a similar virus. Indeed, several authors have argued that the virus is still present in our populace at low levels, or at least that it has popped-up on occasion since its apparent demise in the early 1930's. The other side of this argument has been raised by Mahendra, who has suggested that the apparent decline in catatonic schizophrenia in the latter two-thirds of the 20th century is a reflection of the disappearance of catatonia due to encephalitis lethargica (Mahendra 1981). Extending this idea, a significant portion of the cases of catatonic schizophrenia diagnosed in the past may have been the parkinsonian form of encephalitis lethargica masquerading as a primary mental disorder.

Regardless of whether the virus responsible for encephalitis lethargica has any causative role in schizophrenia today, this unusual viral epidemic holds important heuristic implications for schizophrenia pathogenesis. As stated by McKenna in considering the organic origins of schizophrenia: there is one organic brain disorder "....... which has affiliations with schizophrenic symptoms above all others, and where the relationship is so intimate that it can almost be regarded as schizophrenia's organic next of kin. This is encephalitis lethargica." (McKenna 1994).

3.2.3 Subacute Sclerosing Panencephalitis (SSPE)

SSPE is a degenerative disease of the central nervous system caused by the measles virus. The initial measles infection occurs in early infancy but the patient remains healthy and devoid of any evidence of an insidious disease until many years later when the neuropsychiatric sequela take hold, usually in later childhood or adolescence. Although some of the details of SSPE pathophysiology remained unclear, the prevailing view is that it is caused by a mutant form of the measles virus which is permitted by the host immune system to replicate within the CNS. The virus causes progressive damage as it spreads from the cerebral cortex, to subcortical regions, and ultimately to the brain stem. This process unfolds over months or years, and death usually ensues in one to four years after the onset of neurological symptoms.

The initial clinical manifestations of SSPE involve psychiatric and behavioral alterations. Patients show irritability and affective disturbances, as well as a decline in intellectual functioning that is often noted as deterioration in school performance. As focal neurological signs and dementia set in, the hideous nature of the disease becomes recognized. Several case studies demonstrate that SSPE can produce catatonia and other psychotic symptoms that may be confused with schizophrenia. This is conveyed clearly in the following case report, which was abstracted from Ducalf et. al (Duncalf et al. 1989):

The patient was a 21 year-old female whose infectious disease history was notable for measles at the age of one year. She had no previous psychiatric disorders, and denied use of prescription or illegal drugs. She did not have a history of mental illness in her family. Upon initial presentation, she had delusions of self-reference and persecution, and occasional echolalia. She was not delirious and her sensorium was clear. Her psychotic symptoms had developed over months: she felt people were after her; that someone was tampering with her thoughts; and that radio announcers were talking about her and communicating with police, who were monitoring her with hidden cameras and putting the idea to commit crime into her head. The patient was initially diagnosed with schizophreniform disorder and improved somewhat on fluphenazine, but over the next several months her neurological abnormalities became apparent. These included urinary hesitancy, abnormal gait, partial receptive dysphasia, memory deficits, and myoclonic jerking. Results of an EEG recording and a computerized tomography were consistent with SSPE. The patient became comatose and died several months after her initial presentation. The diagnosis of SSPE was confirmed at autopsy.

Measles virus has been tentatively implicated in multiple sclerosis, and has been considered (with even less certitude) to play a role in some cases of autism (Wakefield et al. 1999). There is currently no evidence to suggest that measles causes schizophrenia. Nevertheless the clinical features of SSPE are instructive for the viral hypothesis of schizophrenia because they demonstrate that psychosis can arise after a long clinically-silent interval between infection and disease onset. Moreover, it is of historical significance that SSPE was first described in 1933 (and perhaps as early as 1912), yet the viral origin of the disease was not established until the advent of modern virological techniques in the late 1960's. SSPE represents yet another example of how viruses can employ perplexing mechanisms to cause psychiatric symptoms in only a subset of infected individuals. The reason why some individuals get SSPE while the vast majority of measles-infected people are spared this neurological complication has not been determined but could involve host factors, environmental influences, or co-infections with other viruses.

3.2.4 HIV Encephalitis

HIV encephalitis is a neurological complication of infection with the human immunodeficiency virus (Navia et al. 1986; Sewell 1996). Although there are two species of HIV (HIV-1 and HIV-2), most of the data on neuropsychiatric complication concerns HIV 1, which is also much more prevalent.

Infection with HIV can result in a plethora of neuropsychiatric sequela affecting every level of the neuroaxis. Accordingly, a substantial lexicon has sprung-up in attempts to describe the various neurological and psychiatric syndromes and complexes attributable to HIV infection. The various shades in meaning of these terms is beyond the scope of our subject, but for the purposes of this chapter the reader should understand that HIV is a lymphotropic and neurotropic virus that causes immunosuppression and allows the invasion and replication of opportunistic pathogens, including many that can infect the central nervous system. The various neuropsychiatric sequela that occur in HIV infection and acquired immunodeficiency syndrome (AIDS) can be caused by replication of the retrovirus within glia of the central nervous system, but can also derive from more indirect means such as the opportunistic infections, metabolic or nutritional complications, and the effects of drugs used to treat the disease. Neuropsychiatric signs and symptoms are common in HIV infection: 30% to 60% of patients experience some central nervous system complication during the course of their illness, and 90% show neuropathological abnormalities at autopsy.

Among psychiatric complications of HIV infection, AIDS dementia complex has been most thoroughly studied. However, there is growing interest in mood and psychotic disorders that emerge as a consequence of HIV infection. One difficulty with studies that endeavor to dissect-out exact causal relationships between HIV infection and the development of psychotic disorders derives from the increased prevalence of HIV in patients with chronic mental illnesses, including schizophrenia. For example, a recent report (which collapsed data across several studies) estimated that the rate of seropositivity for HIV in schizophrenia patients is 9.2%, compared with only 0.4% in the general population (Cournos et al. 1997). Presumably, a substantial portion of this increased infection with HIV stems from risky behavior and other lifestyle variables associated with severe chronic mental illness.

In any case, it is generally agreed that HIV infection can lead to psychotic symptoms itself. Supporting this is the relatively high frequency of new onset psychosis in patients with HIV infection. For example, in one study 15.6% of patients with neurological complications of AIDS had psychosis as an overt clinical feature of the syndrome (Navia et al. 1986). Almost all patients with HIV-associated psychosis have delusional beliefs, which are typically described as being of the paranoid, grandiose, or bizarre type. Auditory and visual hallucinations are also frequent. Psychosis in HIV infection may occur as discreet episodes (45% of cases), or it may be persistent (55% of cases) (Sewell 1996).

HIV is a relatively new infection of humans, having entered our species only within the last 100 years. Therefore, HIV is very unlikely to be the cause of idiopathic schizophrenia. Nevertheless, HIV associated encephalitis is pertinent to the viral hypothesis for schizophrenia for several reasons:

i) HIV is a neurotropic virus that enters the CNS at the time of initial infection, yet the virus has a latency of several years before it produces symptoms. During this latent period, the virus is occult and cannot generally be detected unless highly sensitive PCR techniques are used. Such techniques obviously require considerable knowledge about the viral genome, and prior to their development it was debated as to whether the virus was present in the circulation at all. Thus, the lesson for the viral hypothesis of schizophrenia is that the absence of detectable virus does not imply the absence of a smoldering virus-induced disease process.

ii) Even in cases when opportunistic infections are apparently not present, HIV infection of the brain can cause complex neurobehavioral syndromes such as AIDS dementia. It achieves this without infecting neurons and hence its neuropathogenicity arises from a complex interaction with host molecules and non-neuronal cells.

iii) HIV infection is a neurophysiological and psychological stressor that is likely to exacerbate pre-existing psychotic illnesses. This implies that a virus does not need to be the direct cause of psychotic episodes but rather can act on brain physiology though various pathways to reveal underlying psychiatric vulnerabilities.

iv) HIV- associated psychosis responds to anti-psychotic drug therapy, and may also respond to anti-retroviral medications. This is consistent and the idea that the virus induces a chain of events leading to psychosis, and that antipsychotic drugs can (at least to a degree) interrupt or override this pathological concatenation of events. There is growing interest in studies showing the ability of antipsychotic drugs to have anti-viral properties.

v) There are an estimated 36 million people worldwide who are infected with HIV. If we estimate that 3 percent of these will develop psychosis at some point in their disease, this represents over 1 million psychiatric cases, hence ranking HIV as a significant cause of organic psychosis. The actual increase in psychotic disease burden attributable to HIV is unknown and obviously depends on the prevalence of psychosis in uninfected comparison-populations, as well as the survival period for infected victims. The point with regard to the viral hypothesis of schizophrenia is that even a virus that causes psychosis only rarely can still have a significant impact on the mental health of a population if the virus is sufficiently common.

3.3 THE VALIDITY GAP FOR THE ENCEPHALITIS CONNECTION

The connection between viral encephalitis and psychotic symptoms is intriguing; still, the foregoing clusters of data leave one grappling for a coherent and convincing theory linking neurotropic viruses to idiopathic schizophrenia. The reasons for this unsatisfying void can be considered.

Schizophrenia is associated with negative symptoms (avolition, affective flattening, poverty of speech). These symptoms tend not to be mentioned by proponents of the idea that viruses are a proximal cause of schizophrenia. However, the spectrum of encephalitis-induced symptoms certainly includes loss of motivation, apathy, social withdrawal, poverty of speech, and flattening of affect. Interestingly, negative symptoms are also common in diseases characterized by slow infections prions. In fact, a recent report found that in the majority of patients' psychiatric symptoms including dysphoria, social withdrawal, and apathy, comprised the first symptoms of infection with the new variant Creutzfeldt-Jakob disease supposedly a bovine-transmitted or "Mad-Cow" form of the disease affecting predominately young adults (Spencer et al. 2002). In any case, negative symptoms are notoriously difficult to evaluate and classify. Certainly, a patient with the parkinsonian form of encephalitis lethargica who had avolition, flat affect, and poverty of speech, would not immediately bring to mind the picture of a patient suffering from the negative symptom complex of schizophrenia. It is fair to speculate that the reason negative symptoms have not been emphasized in the proposed connection between viral encephalitis and schizophrenia is that these abstruse symptoms are less likely than florid psychosis to invoke a compelling connection with schizophrenia.

Another reason that information concerning viral encephalitis does not immediately invoke a connection with schizophrenia is that schizophrenia is not diagnosed by a single episode of abrupt-onset psychosis, but rather is a chronic disease that, while having a variable course between individuals, is not usually preceded by an encephalitis-like prodrome. However, as described above, psychotic symptoms can arise before, during, or after a bout with viral encephalitis-- and these symptoms may be episodic or persistent. A viral infection is most often judged to be the culprit in producing the psychosis in instances when the infection is medically severe and occurs in close temporal proximity to the psychotic episode.

A critical evaluation of the proposed connection between encephalitis and schizophrenia raises some additional questions. If encephalitis is responsible for schizophrenia, what information (coming from the fields of neuropsychiatry, neuropathology, and epidemiology) would be required in order to establish this relationship? Could a virus act early to produce a schizotypal syndrome that would become a full-blown psychosis as a consequence of later psychosocial stressors? Conversely, could inherent neurocircuitry vulnerabilities be brought to the surface by a relatively mild neurotropic viral infection?

Because viral encephalitis, as typically defined (e.g. herpes encephalitis), is not likely to go unnoticed clinically (or in a patient's history), these questions bear more on virology then psychiatry. That is, advances in virology determine the fidelity of detection of subclinical viral infections as well as how vestiges of such infections can be discerned. Critics of the viral hypothesis might be persuaded if they saw evidence of infection with a neurotropic virus in a schizophrenia patient's history or through serological testing; and even then, the pathophysiological connection would be far from established. However, at this point, virology does not always offer this level of certitude.

With regard to subclinical viral infections there is currently scant evidence that a virus can invade the brain parenchyma and replicate effectively without producing severe disease. It has been argued that JC virus, which infects approximately 70% of all people, can seed the brain parenchyma and remain latent until reactivated by immunosuppression (Elsner et al. 1992). JC virus causes progressive multifocal leukoencephalopathy in immunouppressed patients. Whether JC or related viruses are connected to schizophrenia remains to be determined, but a role these viruses is not supported by neuropathological examinations. More germane to the viral hypothesis of schizophrenia are numerous attempts to isolate viruses from the brains of patients with schizophrenia and related disorders. These viruses are presumed to have invaded the brain but failed to produce recognizable encephalitis. Although these types of studies are on-going and will continue to employ more sensitive techniques, when considered collectively, the current data do not constitute a credible body of evidence that can consistently point to a viral culprit (Kirch 1993; Yolken et al. 1995). Most recent studies have been negative (Table 1).

Table 1. Evidence of Viral Infection in Brains of Schizophrenia Patients

Viruses[1]	Patients	Test Used	Findings	Reference
HSV-I	1 SCH[4] and no NPC[5]	NAD[2]	SCH was positive	Sequiera (1979)
CMV	6 SCH and 6 NPC	NAD	No positive detections	Aulakh (1981)
HSV-I, CMV, VZV, JCV (PML), BKV	20 SCH and 21 NPC	NAD	No positive detections	Carter (1987)
HSV-I, CMV	25 SCH and 29 NPC	NAD, in vitro	No evidence found	Taylor (1986)
CMV	7 SCH, 5 SCH-like patients, and 9 NPC	NAD	Only 1 SCH exhibited a clear signal	Moises (1988)
Measles, Mumps, Rubella	23 SCH and 23 NPC	NAD	Negative	King (1989)
CMV, etc.[3]	31 SCH and 23 NPC	NAD	Negative in all samples	Taller (1996)
HSV-I, VZV	8 SCH and 8 NPC	NAD	Negative in all samples	Alexander (1992)
HSV-1, HSV-2, CMV	15 SCH and 17 NPC	IHC[4]	Negative in all samples	Stevens (1984)
CMV, HIV, Influenza A, BDV, BVDV	3 SCH and 3 NPC	NAD	Negative in all samples	Sierra-Honigmann (1995)
CMV	25 SCH and 23 NPC	NAD	None detected in any sample	Taylor (1985)

Viruses[1]	Patients	Test Used	Findings	Reference
BDV	16 SCH and 20 NPC	NAD	Negative in all samples	Shankar (1992)
BDV	17 SCH and 10 NPC	NAD	Evidence of virus found in SCH only	Salvatorre (1997)
BDV	9 SCH and 31 NPC	NAD	No significant difference	Haga (1997)
BDV	13 SCH and 52 NPC	NAD	No specific signals detected	Czygan (1999)
BDV	4 SCH and 2 NPC	NAD, IHC, in vivo, in vitro	Presence in CSF/serum of 1 SCH; brain tissue of all SCH positive by some tests	Nakamura (2000)
ERV	4 SCH and 6 NPC	NAD	45% of clones obtained from SCH were homologous to MSRV and 10% from NPC	Yolken (2000)
Retrovirus	5 SCH and 6 NPC	NAD	Retroviral sequences found more often in CSF of SCH than in NPC. An MSRV-like sequence was upregulated in cortex of SCH.	Karlsson (2001)

Viruses[1]	Patients	Test Used	Findings	Reference
VZV, HSV, CMV	6 SCH, 2 equivocal SCH, 6 non-schizophrenic psychiatric controls, and 6 NPC	IHC	4 SCH and both questionable SCH showed signs of VZV and CMV along with 4 NPC; no HSV evidence found	Stevens (1983)

[1] BDV = Borna Disease Virus; BKV = Papovirus BK Virus; BVDV = Bovine Viral Diarrhea Virus; CMV = Cytomegalovirus; EBV = Epstein-Barr Virus; ERV = Endogenenous Retrovirus; HHV = Human Herpesvirus; HIV = Human Immunodeficiency Virus; HSV = Herpes Simplex Virus; JCV = Human Polyoma Virus JC; MSRV=multiple sclerosis retrovirus; PML = Progressive Multifocal Leukoencephalopathy; VZV = Varicella – Zoster Virus

[2] NAD = Nucleic Acid Detection such as southern blots, northern blots, RT-PCR, dot blots, etc.

[3] Also tested for EBV, HSV-I, VZV, HHV-6, Measles, Mumps, Rubella, Picornavirus, Influenza A, HTLV-I, and St. Louis Encephalitis virus, all with negative results.

[4] IHC = Immunohistochemistry; also includes western blots, dot blots with antibodies, etc.

[4] SCH = Schizophrenia Cases

[5] NPC = Non-psychiatric Controls

In any case, viral encephalitis can masquerade as schizophrenia, and Torrey has supplied a practical set of guidelines for the clinician to decide " ...when to call a virologist to the psychiatric ward" (Torrey 1986). He suggests that the following conditions should alert the psychiatrist that a neurological or infectious disease work-up might be needed:

i) Patient presents with immediately preceding or concurrent signs or symptoms of "flu", upper respiratory infection, fever, or headaches.

ii) Extremely disinhibited behavior (e.g. public masturbation) and/or bizarre delusions (e.g. man with influenza encephalitis who believed he was a gorilla).

iii) Rapid onset of the psychosis.

iv) Seizure-like activity, vague neurological symptoms, or neurological signs such as ataxia, nystagmus, ocular palsies, or pupillary changes.

v) Olfactory or visual hallucinations, rapid fluctuations in mental status, or other signs commonly attributed to organic causes of psychosis.

Schizophrenia is a heterogeneous disorder. Taking this into consideration it will be difficult to determine how many patients who carry the diagnosis of schizophrenia actually have an encephalitis-associated disease; this number is, however, almost certainly not zero.

3.4 BORNA DISEASE VIRUS.

Presently, the most controversial issue in neurovirology is the role of Borna disease virus (BDV) in human neuropsychiatric disorders. Intensifying the controversy, the researchers involved are generally top-level neurovirologists and immunologists who have impressive track records in discerning mechanisms of viral pathogenesis (Allmang et al. 2001; Amsterdam et al. 1985; Bode et al. 1992; Carbone 2001; Chen et al. 1999; Czygan et al. 1999; Nakamura et al. 2000; Richt et al. 1997; Rott et al. 1985; Salvatore et al. 1997b; Schwemmle et al. 1999; Staeheli et al. 2001; Stitz et al. 2002; Waltrip et al. 1995). Thus the controversy cannot be dismissed as foggy thinking or sloppy experimentation on either side.

There is a range of views concerning the role of BDV in brain diseases; and while the extremes at either pole of this issue can be delineated, few individuals actually hold these extreme views. The extreme BDV advocates would argue that the virus is an endemic neurotropic pathogen that plays a causative role in a myriad of maladies including major depression, bipolar disorder, schizophrenia, multiple sclerosis, non-Alzheimer's dementia, chronic fatigue syndrome, motor neuron disease, and glioblastoma. The opposing view is antithetical to any involvement of BDV in human disease. The contention behind this view is that BDV does not even infect humans, and worse yet, the advocates for BDV have been chasing a series of laboratory artifacts.

The rationale for investigating a potential role of BDV in neuropsychiatric disorders is rock-solid. BDV is a negative sense RNA virus that is remarkable for its unique replication strategy and extremely broad host range. BDV causes natural infection in horses, sheep, cattle, dogs, cats, and birds. In addition, experimental infections have been established in a plethora of other species including rodents and nonhuman primates.

Experiments performed over the last 70 years have established beyond a doubt that BDV can infect the nervous system of animals and cause neuropathological and behavioral alterations. The virus targets the limbic system, and in rodents it has been reported to impact neurotransmitter systems including dopamine, neuropeptides, and glutamate (Gosztonyi 2001; Hornig et al. 1999). The evidence that BDV causes neuropsychiatric disorders in humans has accumulated over the last 17 years. All of it is controversial.

Before considering the vast array of studies pointing to a causative role for BDV in neuropsychiatric disorders, it is helpful to consider what types of evidence would convince even the most strident critics. If BDV is involved in schizophrenia pathogenesis, one could construct a hypothetical scenario based on the virus's known properties in animals, and characteristics of other neural infections in humans. It is important to emphasize that the following is just a hypothetical scenario to aid the reader in scrutinizing the actual data on BDV in humans.

Presumably, the virus would enter the human victim through the respiratory route. The infection could be acquired at any time during postnatal brain development, or during adulthood. Epidemiologically, one would expect patients with schizophrenia to have had increased contact with natural sources of the virus; working on a farm for example since livestock seem to be the most robust source. Alternatively, patients could have a genetic susceptibility that made them more vulnerable to infection whereas people without schizophrenia would be resistant. The virus, like most pathogens, would be expected to provoke an immune response detectable in plasma as a transient rise in IgM followed by a switch to IgG antibodies that have high affinity binding to multiple viral proteins. The virus would enter the CNS, either by tracking along olfactory nerves or via the hematogenous route. In adults, an inflammatory encephalitis would be expected, with focal neurological signs. After the acute viral phase subsided, chronic behavioral symptoms would set in. Since it is a persistent virus, BDV could be isolated from white blood cells in most cases, and could be detected by RT-PCR in virtually all cases. The genomic sequence of the virus would reflect its human origin by differing from laboratory strains. Viral proteins could be detected in white blood cells by immunostaining with anti- BDV antibodies. Patients without schizophrenia would (under this scenario) be negative for all these tests. Even the critics would not expect that the virus could be isolated from the plasma of schizophrenia patients; the critics would also allow that some infected patients may not have enough virus in their white blood cells to be isolated in the laboratory. In this idealized scenario, autopsy would be definitive--which is asking a lot in the case of schizophrenia. BDV antigens and nucleic acids would be detectable in brain regions affected by schizophrenia, but only in patients, and not in controls. Inoculation of brain material from patients with schizophrenia into nonhuman primates would produce a complex behavioral syndrome (even the critics would not expect schizophrenia in a monkey).

Now, let us consider the actual data related to BDV in humans. Epidemiological data on transmission from farm animals to humans is incomplete but a recent study indicates that people living near horse farms are more likely to have serological and RT-PCR evidence of BDV infection than controls (Takahashi et al. 1997). But schizophrenia is associated with birth and domicile in urban as opposed to rural areas, so this does not fit with a BDV etiology of schizophrenia.

Assuming that BDV does infect humans, the symptoms produced by even a peripheral infection are unknown. Therefore, serological studies of acute (IgM) titers after BDV infection are lacking. However, in 1985 it was reported that a small but significant portion of psychiatric patients, but not controls, had serum antibodies against BDV (Amsterdam et al. 1985; Rott et al. 1985). Subsequently, there has been a steady stream of papers entering the literature that have examined BDV serology in neurological and psychiatric patient populations. Strangely, a variety of seemingly unrelated disorders of the brain have been reported to be associated with circulating antibodies against BDV. Depending on the techniques used and the criteria for seropositivity, the percentage of the schizophrenia patients having detectable anti- BDV antibodies ranges from 0 to 32.2% (Kubo et al. 1997; Waltrip et al. 1997; Waltrip et al. 1995). However, seropositivity in psychiatric control populations ranges from 0 to 20%.

Supporting the claims of the BDV proponents, a variety of immunological detection techniques have been used successfully to measure anti-BDV antibodies in patient' sera. These techniques vary in the degree to which the antigen is denatured, and hence positive findings using an assay to detect denatured (linear) epitopes are complemented by studies using techniques that are more likely to leave comformational epitopes intact.

Nevertheless, the low avidity of anti-BDV IgG in most psychiatric patients, and the failure of these to undergo "affinity maturation" as would be expected in an antiviral immune response, has called into question whether the antibodies detected in human serum are actually directed against BDV components. Moreover, even when the same serum samples are assayed by two different methods (e.g. ELISA and immunoblot), a substantial portion of patients may be positive by one method and negative by the other. In contrast to serological studies in experimental animals, most patients have detectable antibodies to only one of the major BDV antigens. Still, some patients' sera do react with more than one BDV protein, and when the serological data are judged based on this more stringent criteria, the differences between schizophrenia patients and controls in antibody prevalence are greatly enhanced (Waltrip et al. 1995)

The precise reason for the great variability between studies has not been deciphered, but one confounding factor is that BDV antigens may circulate as immune complexes bound to the P24 and P40 proteins of the virus. This implies, of course, that there are viral antigens in the patient's serum. Recently, an assay to measure these immune complexes has been developed. Preliminary results using this assay indicate that about 25% of normal control patients have such evidence of BDV infection, but fully 100% of patients with major depression carried this marker (Bode et al. 2001).

Overall, the bulk of evidence indicates that a subset of psychiatric patients are more likely than mentally--healthy controls to have immunological indices of infection with BDV or a related agent. However, discrepancies in the literature are abundant, and the debate is certainly not settled.

If BDV causes schizophrenia, it presumably does so by entering the brain. Is there any evidence for BDV- induced encephalitis in humans? As discussed previously, psychiatric symptoms can arise as a consequence of viral encephalitis, and often the causative pathogen is not identified. The techniques for detecting BDV in the brain or CSF have just recently been refined to the point that this question can now hopefully be addressed. At present, it is unknown whether BDV causes clinical or subclinical encephalitis in humans.

BDV has low cytopathogenicity and has characteristics of persistent virus. Attempts to isolate replicating BDV from peripheral blood have been successful in a few cases but the similarity of these isolates to laboratory strains has raised the specter that the putative isolates may be contaminants (Schwemmle et al. 1999). A number of groups have also detected BDV in peripheral blood cells from psychiatric patients at much higher frequency than controls. As with the serology tests, positive findings have been reported not just in schizophrenia but also in a variety of other neuropsychiatric disorders. In one study, 7 out of 11 schizophrenia patients were positive for BDV RNA by RT-PCR, compared with none of the 23 controls (Sauder et al. 1996). While several other studies have found evidence of BDV proteins or RNA in peripheral blood mononuclear cells of schizophrenia patients, but not controls, these findings are still controversial. Some studies were unable to find evidence of BDV in peripheral blood mononuclear cells despite the use of highly-sensitive RT-PCR methods. This could be because these studies were looking in the wrong place; that is, the virus may be harbored in granulocytes, which might only be present in PBMC fractions at the low levels (Planz et al. 1999).

The issue of contamination continues to hang over BDV studies using RT-PCR methods. In a collaborative effort to address this issue, the same samples were analyzed in three different labs (Richt et al. 1997). Only the laboratory presently engaged in BDV work (and thus subject to contamination) yielded positive results in any of the psychiatric patients tested.

BDV protein nucleic acids have been detected in postmortem studies of patients with neurological or psychiatric disorders, including schizophrenia. In one study, RT-PCR gave a positive result in 9 out of 17 schizophrenia patients, and in zero of 10 controls who did not have a neuropsychiatric disorder (Salvatore et al. 1997a). In another study, the presence of BDV

RNA was analyzed by both RT-PCR and in situ hybridization in 13 brain regions of one patient who died of a myocardial infarction at 26 years old (Nakamura et al. 2000). Amplification of BDV RNA was found in samples of his temporal lobe, hippocampus, pons, and cerebellum. The signal in the hippocampus, pons, and cerebellum was confirmed by in situ hybridization, and showed a pattern suggestive of neuronal infection. Experimental infection of a gerbil with brain homogenate from this patient was apparently successful. While these data are intriguing, the problem of contamination continues to loom. The genetic sequence of the virus infecting this person appears usually similar to laboratory strains (Staeheli & Lieb 2001). This normally raises a red flag to virologists, who expect some genetic drift in a virus as it infects new hosts. But BDV appears to be an evolutionarily-old virus with a surprisingly stable genome, so this concern may not apply to BDV.

At face value, the Borna disease theory for schizophrenia seems to defy a rational model of pathogenesis. Why is the virus present in so many neurological and psychiatric disorders? What can account for the discrepancies between different studies, and the mismatched indices of infection within a single case (e.g. a patient may be positive for antibody but negative for PBMC infection by protein anaysis or RT-PCR)? If the virus is supposedly persistent, why can't it be detected in all patients who are supposedly infected?

One possibility is that BDV infection of psychiatric patients is an epiphenomena. Major depression, bipolar disorder, and schizophrenia are well known to be associated with immune changes. It is inconceivable that a selective immunosuppression in some patients due to mental illness opens an opportunity for BDV to infect one of its unusual hosts: in this case human beings. Another possibility is that BDV is a nosocomial infection of psychiatric care facilities. Chen et. al examined the prevalence of BDV RNA in PBMCs from schizophrenia patients, normal controls, and mental-health care workers (psychiatrists, nurses, social workers, and occupational therapists). In the control (community sample) only 1.4% were positive for BDV, but among schizophrenia patients, 13.5% were positive (Chen et al. 1999). Surprisingly, the highest rate of apparent infection (15%) was found in the mental health care workers who did not have serious mental illness but who were in contact with inpatient psychiatric patients.

The question of whether BDV or a related virus can cause schizophrenia remains unsettled. BDV appears to be a virus that has adapted through evolution to many different hosts, particularly hoofed animals. BDV may have entered the human population coincident with domestication and the close association with these animals. If BDV causes schizophrenia, this would imply that schizophrenia may not be as ancient as previously thought; a point which has been raised frequently by several epidemiologists.

In any case, all indications point to the ability of BDV to infect virtually every warm blooded animal. Considering the long-time physical proximity of humans with feral and domesticated animals, it seems likely that humans have been and will continue to be infected with BDV. It also seems wishful thinking to conclude that every member of our human race has been and will continue to be spared any deleterious consequences of infection with this highly neurotropic virus.

3.5 VIRUSES IN OTHER PSYCHIATRIC DISEASES:

Viruses have been tentatively implicated in most major psychiatric diseases of unknown etiology. The contribution of infectious diseases to the pathophysiology of major depression and bipolar disorders has long been considered (Yolken & Torrey 1995). As mentioned above, there has been renewed interest in this area based on the isolation of Borna disease virus from the immune cells of patients with depression. One of the prevailing theories of how viruses may be involved in major depression involves immune activation. For example, elevated serum concentrations of the proinflammatory cytokines IL-1 and IL-6, as well as acute phase proteins, have been found in patients with major depression and may indicate the presence of a smoldering infectious disease process (Maes 1999). There has been considerable interest in the role of viral infections in bipolar disorder. Many of the concepts and techniques that have been applied to the viral hypothesis of schizophrenia have also been applied to bipolar disorder. In addition, Alzheimer's disease has been proposed to be caused by a virus that perhaps enters through the nasal mucosa and infects the olfactory bulb before tracking to other brain regions. While this remains speculative, there is strong evidence to indicate that the immune system contributes to Alzheimer's pathogenesis (Benveniste et al. 2001). Chronic fatigue syndrome is an illness with a controversial etiology and pathogenesis. Tests of immune function have found indications of both immune activation and immunosuppression. An acute viral illness frequently preceeds the onset of chronic fatigue syndrome yet no infectious agent has been consistently identified as causing the disease.

In contrast to the unclear relationship of viruses to these neuropsychiatric disorders, bacteria are well-established causative agents of psychiatric symptoms. Besides the obvious example of syphilis, which is well known to cause psychosis and a variety of other neuropsychiatric sequela, the tick-borne bacteria responsible for Lyme disease is also remarkable because of the spectrum of psychological abnormalities it induces including anxiety, irritability, obsessions and compulsions, and cognitive deficits.

Although it is well recognized that viruses can bring on sickness behavior, which has all the stigmata of depression, and most clinicians are familiar with instances where a bout of major depression was initially triggered by an evanescent viral infection, there is still no firm evidence that viruses are a significant cause of any of common idiopathic psychiatric disorders (Miller et al. 1995). More research in this area is definitely needed. Since there are likely to be overlapping pathophysiological mechanisms for most psychiatric disorders, by discerning the role of infections and immune processes in one disorder, we may gain clues to the pathophysiology of the others.

REFERENCES

Allmang U, Hofer M, Herzog S et al. Low avidity of human serum antibodies for Borna disease virus antigens questions their diagnostic value. Molecular Psychiatry 2001; 6:329-33.

Alexander RC, Cabirac G, Lowenkopf T et al. Search for evidence of herpes simplex virus, type 1, or varicella-zoster virus infection in postmortem brain tissue from schizophrenic patients. Acta Psychiatrica Scandinavica 1992; 86:418-20.

Amsterdam JD, Winokur A, Dyson W et al. Borna disease virus. A possible etiologic factor in human affective disorders? Archives of General Psychiatry 1985; 42:1093-6.

Aulakh GS, Kleinman JE, Aulakh HS et al. Search for cytomegalovirus in schizophrenic brain tissue. Proceedings of the Society for Experimental Biology & Medicine 1981; 167:172-4.

Benveniste EN, Nguyen VT, O'Keefe GM. Immunological aspects of microglia: relevance to Alzheimer's disease. Neurochemistry International 2001; 39:381-91.

Bode L, Reckwald P, Severus WE et al. Borna disease virus-specific circulating immune complexes, antigenemia, and free antibodies--the key marker triplet determining infection and prevailing in severe mood disorders. Molecular Psychiatry 2001; 6:481-91.

Bode L, Riegel S, Lange W et al. Human infections with Borna disease virus: seroprevalence in patients with chronic diseases and healthy individuals. Journal of Medical Virology 1992; 36:309-15.

Carbone KM. Borna disease virus and human disease. Clinical Microbiology Reviews 2001; 14:513-27.

Carter GI, Taylor GR, Crow TJ. Search for viral nucleic acid sequences in the post mortem brains of patients with schizophrenia and individuals who have committed suicide. Journal of Neurology, Neurosurgery & Psychiatry 1987; 50:247-51.

Chen CH, Chiu YL, Shaw CK et al. Detection of Borna disease virus RNA from peripheral blood cells in schizophrenic patients and mental health workers. Molecular Psychiatry 1999; 4:566-71.

Cournos F, McKinnon K. HIV seroprevalence among people with severe mental illness in the United States: A critical review. Clinical Psychology Review 1997; 17:259-69.

Czygan M, Hallensleben W, Hofer M et al. Borna disease virus in human brains with a rare form of hippocampal degeneration but not in brains of patients with common neuropsychiatric disorders. [see comments.]. Journal of Infectious Diseases 1999; 180:1695-9.

Duncalf CM, Kent JNG, Harboard M et al. Subacute sclerosing panencephalitis presenting as schizophreniform psychosis. British Journal of Psychiatry 1989; 155:557-9.

Elsner C, Dorries K. Evidence of human polyomavirus BK and JC infection in normal brain tissue. Virology 1992; 191:72-80.

Gosztonyi G, editor The mechanism of neuronal damage in virus infections of the nervous system Heidelberg: Springer; 2001. (Current Topics In Microbiology and Immunology).

Haga S, Motoi Y, Ikeda K. Borna disease virus and neuropsychiatric disorders. Lancet 1997; 350:592-3.

Hornig M, Weissenbock H, Horscroft N et al. An infection-based model of neurodevelopmental damage. Proceedings of the National Academy of Sciences of the United States of America 1999; 96:12102-7.

Jennings LC, Carrasco LH, Lord MA et al. Polymerase chain reaction (PCR) search for viral nucleic acid sequences in schizophrenia. Lancet 1979; 2:609-12.

Karlsson H, Bachmann S, Schroder J et al. Retroviral RNA identified in the cerebrospinal fluids and brains of individuals with schizophrenia. [see comments.]. Proceedings of the National Academy of Sciences of the United States of America 2001; 98:4634-9

Kirch DG. Infection and autoimmunity as etiologic factors in schizophrenia: a review and reappraisal. Schizophrenia Bulletin 1993; 19:355-70.

King DJ, Bachmann S, Schroder J et al. Viruses, immunity and mental disorder. Proceedings of the National Academy of Sciences of the United States of America 2001; 98:4634-9.

Kubo K, Fujiyoshi T, Yokoyama MM et al. Lack of association of Borna disease virus and human T-cell leukemia virus type 1 infections with psychiatric disorders among Japanese patients. Clinical & Diagnostic Laboratory Immunology 1997; 4:189-94.

Lewis D. Retroviruses and the pathogenesis of schizophrenia. Proceedings of the National Academy of Sciences of the United States of America 2001; 98:4293-4.

Lishman WA. Organic Psychiatry. Oxford: Blackwell; 1987.

Little JT, Sunderland T. Psychosis secondary to encephalitis and ecephalopathies. Seminars in Clinical Neuropsychiatry 1998; 3:4-11.

Maes M. Major depression and activation of the inflammatory response system. Advances in Experimental Medicine & Biology 1999; 461:25-46.

Mahendra B. Where have all the catatonics gone? Psychological Medicine 1981; 11:669-71.

McGrath N, Anderson NE, Croxson MC et al. Herpes simplex encephalitis treated with acyclovir: diagnosis and long term outcome. Journal of Neurology, Neurosurgery & Psychiatry 1997; 63:321-6.

McKenna PJ. Schizophrenia and related syndromes. New York; 1994. pp 418 (quote on page 334) p.

Miller A, Spencer R. Immune system and central nervous system interactions. In: Wa Wilkins, editor Comprehensive Textbook of Psychiatry. Vol. VI. Baltimore; 1995.

Moises HW, Ruger R, Reynolds GP et al. Human cytomegalovirus DNA in the temporal cortex of a schizophrenic patient. European Archives of Psychiatry & Neurological Sciences 1988; 238:110-3.

Morzunov S, Schwemmle M, Lipkin WI et al. Detection of herpes-simplex viral genome in brain tissue. Lancet 1997; 349:1813-4.

Nahmias AJ, Roizman B. Infection with herpes-simplex viruses 1 and 2. 1. New England Journal of Medicine 1973; 289:667-74.

Nakamura Y, Takahashi H, Shoya Y et al. Isolation of Borna disease virus from human brain tissue. Journal of Virology 2000; 74:4601-11.

Navia BA, Jordan BD, Price RW. The AIDS dementia complex: I. Clinical features. Annals of Neurology 1986; 19:517-24.

Planz O, Rentzsch C, Batra A et al. Pathogenesis of borna disease virus: granulocyte fractions of psychiatric patients harbor infectious virus in the absence of antiviral antibodies. Journal of Virology 1999; 73:6251-6.

Richt JA, Alexander RC, Herzog S et al. Failure to detect Borna disease virus infection in peripheral blood leukocytes from humans with psychiatric disorders. Journal of Neurovirology 1997; 3:174-8.

Rott R, Herzog S, Fleischer B et al. Detection of serum antibodies to Borna disease virus in patients with psychiatric disorders. Science 1985; 228:755-6.

Salvatore M, Morzunov S, Schwemmle M et al. Borna disease virus in brains of North American and European people with schizophrenia and bipolar disorder. Bornavirus Study Group. [see comments.]. Lancet 1997a; 349:1813-4.

Salvatore M, Morzunov S, Schwemmle M et al. Borna disease virus in brains of North American and European people with schizophrenia and bipolar disorder. Lancet 1997b; 349:1813-4.

Sauder C, Muller A, Cubitt B et al. Detection of Borna disease virus (BDV) antibodies and BDV RNA in psychiatric patients: evidence for high sequence conservation of human blood-derived BDV RNA. Journal of Virology 1996; 70:7713-24.

Schwemmle M, Jehle C, Formella S et al. Sequence similarities between human bornavirus isolates and laboratory strains question human origin. [see comments.]. Lancet 1999; 354:1973-4.

Sewell DD. Schizophrenia and HIV. Schizophrenia Bulletin 1996; 22:465-73.

Shankar V, Dietzschold B, Fu ZF et al. A search for borna virus-specific genome in human nerve tisssue by reverse transciptase polymerase chaing reaction (RT-PCR). Biological Psychiatry 1992; 31:103A.

Spencer MD, Knight RS, Will RG. First hundred cases of variant Creutzfeldt-Jakob disease: retrospective case note review of early psychiatric and neurological features. BMJ 2002; 324:1479-82.

Staeheli P, Lieb K. Bornavirus and psychiatric disorders -- fact or fiction? Journal of Medical Microbiology 2001; 50:579-81.

Stevens J, Langloss J, Albrecht P et al. A search for Cytomegalovirus and Herpes Viral Antigen in Brains of Schizophrenic Patients. Archives of General Psychiatry 1984; 41:795-801.

Stitz L, Bilzer T, Planz O. The immunopathogenesis of Borna disease virus infection. Frontiers in Bioscience 2002; 7:d541-55.

Takahashi H, Nakaya T, Nakamura Y et al. Higher prevalence of Borna disease virus infection in blood donors living near thoroughbred horse farms. Journal of Medical Virology 1997; 52:330-5.

Taller AM, Asher DM, Pomeroy KL et al. Search for viral nucleic acid sequences in brain tissues of patients with schizophrenia using nested polymerase chain reaction. Archives of General Psychiatry 1996; 53:32-40.

Taylor GR, Crow TJ. Viruses in human brains: a search for cytomegalovirus and herpes virus 1 DNA in necropsy tissue from normal and neuropsychiatric cases. Psychological Medicine 1986; 16:289-95.

Taylor GR, Crow TJ, Higgins T et al. Search for cytomegalovirus in postmortem brain tissue from patients with Huntington's chorea and other psychiatric disease by molecular hybridization using cloned DNA. Journal of Neuropathology & Experimental Neurology 1985; 44:176-84.

Torrey EF. Functional Psychosis and Viral Encephalitis. Intergrated Psychiatry 1986; 4:224-36

Torrey EF. Stalking the schizovirus. Schizophrenia Bulletin 1988; 14:223-9.

Tsuang MT, Faraone SV. The case for heterogeneity in the etiology of schizophrenia. Schizophrenia Research 1995; 17:161-75.

Wakefield AJ, Montgomery SM. Autism, viral infection and measles-mumps-rubella vaccination. [letter; comment.]. Israel Medical Association Journal: Imaj 1999; 1:183-7.

Waltrip RW, 2nd, Buchanan RW, Carpenter WT, Jr. et al. Borna disease virus antibodies and the deficit syndrome of schizophrenia. Schizophrenia Research 1997; 23:253-7.

Waltrip RW, 2nd, Buchanan RW, Summerfelt A et al. Borna disease virus and schizophrenia. Psychiatry Research 1995; 56:33-44.

Whitley RJ.Herpes simplex virus. In: WM Scheld; RJ Whitley; DT Durack, editors Infections of the Central Nervous System. Philadelphia: Lippencott-Raven; 1997.

Wilson LG. Viral encephalopathy mimicking functional psychosis. American Journal of Psychiatry 1976; 133:165-70.

Yolken RH, Torrey EF. Viruses, schizophrenia, and bipolar disorder. Clinical Microbiology Reviews 1995; 8:131-45.

Yolken RH, Karlsson H, Yee F et al. Endogenous retroviruses and schizophrenia. Brain Research - Brain Research Reviews 2000; 31:193-9.

Chapter 4

EPIDEMIOLOGICAL EVIDENCE

The first section of this chapter, describes the epidemiological evidence pertaining to the viral hypothesis of schizophrenia. Some of the studies reviewed in this section have not rigorously applied systematic sampling methods, or they have relied on samples of convenience. Thus, the term "epidemiological evidence" has been applied in its broadest sense. The second section of this chapter, describes evidence that an autoimmune process is involved in schizophrenia. While it could be argued that this is tangential to the viral hypothesis, there is a growing consensus that viruses are capable of triggering autoimmune diseases in humans, making the subject of immune changes in schizophrenia topical to this book.

4.1 EPIDEMIOLOGICAL STUDIES OF THE VIRAL HYPOTHESIS

Because viral infections are often self-limiting, specialists in infectious disease epidemiology are often concerned with "disease incidence": that is, the number of new cases occurring over a limited period of time. This statistic is important for tracking epidemics and establishing that a disease is indeed infectious. For chronic diseases such as schizophrenia, which are not usually associated with dramatic shifts in incidence over time, "prevalence" is often measured. Prevalence is the portion of the population who has a disease at a given time. Incidence and prevalence are related to each other by the number of people who are cured or die (figure 1).

On a fundamental level, the question posed by epidemiological explorations of the viral hypothesis is this: does infection with a virus (or class of viruses) increase a person's risk of developing schizophrenia, and does removing this risk factor (e.g. by prevention) decrease the incidence of schizophrenia.

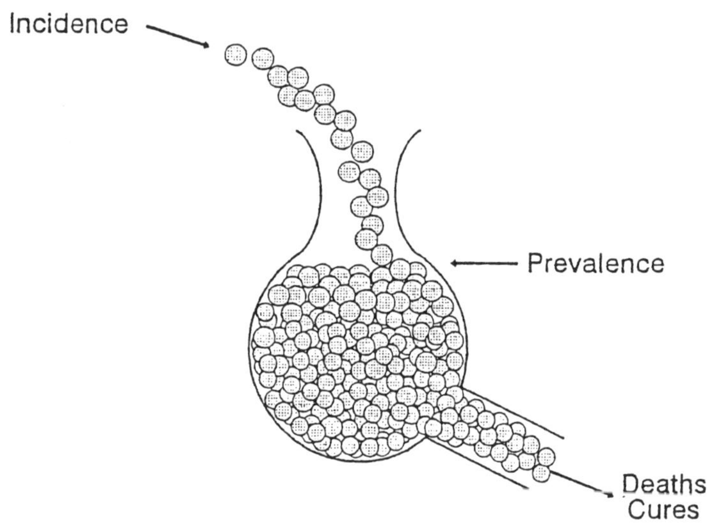

Figure 1 Relationship between incidence and prevalence. Reproduced with permission from: Gordis, L., ed. (2000) Epidemiology, 2nd edition (pp. 35). Philadelphia: W.B. Saunders Company.

The endeavor of establishing risk factors and causal relationships through epidemiological investigations is a difficult and arduous process. In some instances a causal relationship can be established definitively. For example exposure to sunlight is a risk factor for sunburn; and while there are modulating variables such as skin tone and prior exposure, there is no doubt about this connection because the risk factor (sunlight) is visible to the eye, and there is a consistent and close temporal relationship between this factor and the physiological variable (sunburn). The situation with virus-induced chronic diseases is the opposite. Viruses are invisible, and there is a variable and often protracted period between exposure and manifestation of the chronic disease state.

Table 1

Variable or Risk Factor Found in Schizophrenia Patients	Putative Implications for a Viral Etiopathology of Schizophrenia
↑ Exposure to influenza epidemics during 2nd trimester in utero	Maternal influenza (or ensuing immune response) causing disrupted fetal brain development
↑ Prenatal or postnatal exposure to enteroviruses (e.g. Poliovirus)	Subsets of neurons destroyed by neuro-invasive viral strains in susceptible hosts
↑ Incidence of urban versus rural birth, ↑ household crowding	Higher exposure to communicable viruses
↑ Number of siblings or ↑ probability of having older siblings	Transmission of (community acquired) virus to pregnant mother or younger (affected) sibling
↑ Probability of being born in late winter or early spring	Exposure to viral epidemics with seasonal cyclicality, indoors crowding
↑ Schizophrenia incidence among immigrant populations (including 2nd generation)	Immigrants contracting primary infections by novel viruses (e.g. fetus not protected by maternal antibodies)
↑ Exposure in utero to famine	Nutritional deprivation enhancing host vulnerability; heightened viral contact from atypical food sources
↑ Probability of dermatoglyphic or minor physical abnormalities	Altered morphogenesis by prenatal viral infection
↑ Probability of having been born with perinatal or obstetrical complications (OCs)	Virus-induced fetopathy causing OCs; heightened exposure of neonate to nosocomial viral infections
↑ viral antibodies (in serum or CSF) to herpes viruses, influenza A, vaccinia, and others	Evidence of infection, virus reactivation, or virus-induced autoimmunity.

↑--Increased versus Controls

For example, infection with the hepatitis B virus is a major risk factor for developing hepatocellular carcinoma. The magnitude of the effect it is not subtle--people infected with the virus are 100 fold more likely to develop this cancer (Beasley 1988). But this relationship has only recently been established even though the existence of this hepatotropic virus has been known since the 1930's. Part of the difficulty in establishing a viral cause for hepatocellular carcinoma is the variable latency between initial infection and the development of the cancer; a period which may be in excess of 50 years. Moreover, the majority of people infected with the virus never develop the cancer. Besides genetic variables, there are modulating environmental agents such as toxic substances present in ingested fungi that help determine which individuals develop this chronic complication of hepatitis B virus.

To put this into perspective, even the most robust risk factor for schizophrenia (family history) does not increase the odds of getting the disease by 100 fold (Jablensky 1999; Tsuang 2000). Thus our expectations of the epidemiological studies related to the viral hypothesis for schizophrenia should be modest.

Some of the epidemiological studies pertinent to a viral cause of schizophrenia are listed in table 1. Many of these studies did not set out to examine the role of viruses in schizophrenia per se, but their results can be interpreted in light of the viral hypothesis. This is not to imply that a viral etiopathogenesis is necessarily the most parsimonious explanation for the data. A case in point is the increased risk of schizophrenia associated with maternal exposure to famine while in utero (Susser et al. 1992). The simplest explanation for this is that the developing brain is deprived of nutrients and that this brings about alterations that eventually give rise to schizophrenia. But it cannot be denied that famine is associated with a weakening of the immune system and the consumption of unusual food sources, which might increase risks of viral infection. The next section examines specific epidemiological findings in schizophrenia and their proposed connection with viral infections.

4.1.1 Heredity

There are inarguably genetic risk factors for schizophrenia; though the genes involved have not yet been identified (Tsuang 2000). Figure 2 shows the relative risk for developing schizophrenia as a function of shared genes with family members. Clearly having a close relative with schizophrenia is a risk factor for developing the disease. This is illustrated starkly by studies of twins, whereby having and identical twin with schizophrenia increases the risk for developing the disease by 50 fold. But twin studies argue equally for the importance of environmental factors (including infections) in schizophrenia. Monozygotic twins share 100% of their genes but have a concordance rate of only 48%, the remainder being due to environmental influences. Even this figure of 48% may overestimate the genetic contribution to schizophrenia. Depending on the study, and the method used to determine concordance, lower values may be obtained.

Torrey reviewed several studies that used the pairwise method to determine concordance (as opposed to the proband method reported above) and found concordance rates of 28% for monozygotic twins, compared to 6% for dizygotic twins (Torrey 1992). In interpreting data such as that shown in figure 2, it's important to realize that disease rates in families cannot be used as a stand-in for genetic variables. Families tend to have in common the same environment, which includes shared space, toxicant exposure, and contacts with family friends. In the case of twins, there is even a shared uterine environment, as well as similarities in the timing and exposure to postnatal infectious diseases (Torrey et al. 1999).

While the genetic contribution to schizophrenia is undeniable, attempts to view schizophrenia as principally a genetic disease raises more conundrums than solutions. There is the longstanding question of how schizophrenia, which generally decreases the fecundity of those affected, is able to persist for generation after generation. This enigma deepens when one considers the impact of institutionalization patterns on disease prevalence. Whereas institutionalization would be expected to decrease procreation of patients, schizophrenia incidence appears to be decreasing as fewer people are institutionalized. As pointed out by Torrey (Torrey & Yolken 1999) : "If schizophrenia is a truly genetic disease, therefore, it is the only known genetic disease that becomes more common with less reproduction by those affected and less common with more reproduction of those affected" .

In fairness, it must be conceded that psychiatric geneticists have used established methods to estimate schizophrenia heritability, the portion of the phenotype can be explained by genetic variance, and have yielded values ranging from 41% to 87%, with more recent figures edging toward upper end of this range (Cardno et al. 1999). These estimates complement a variety of studies suggesting an important genetic influence on complex behavioral traits-- as indicated in table 2, which shows the results of twin studies for several psychiatric disorders (McGue et al. 1998). Importantly, the psychiatric community has come to the realization with regard to these disorders (as well as schizophrenia) that interventions at the level of the environment are likely to be more easily achieved than attempts to control or alter genotype.

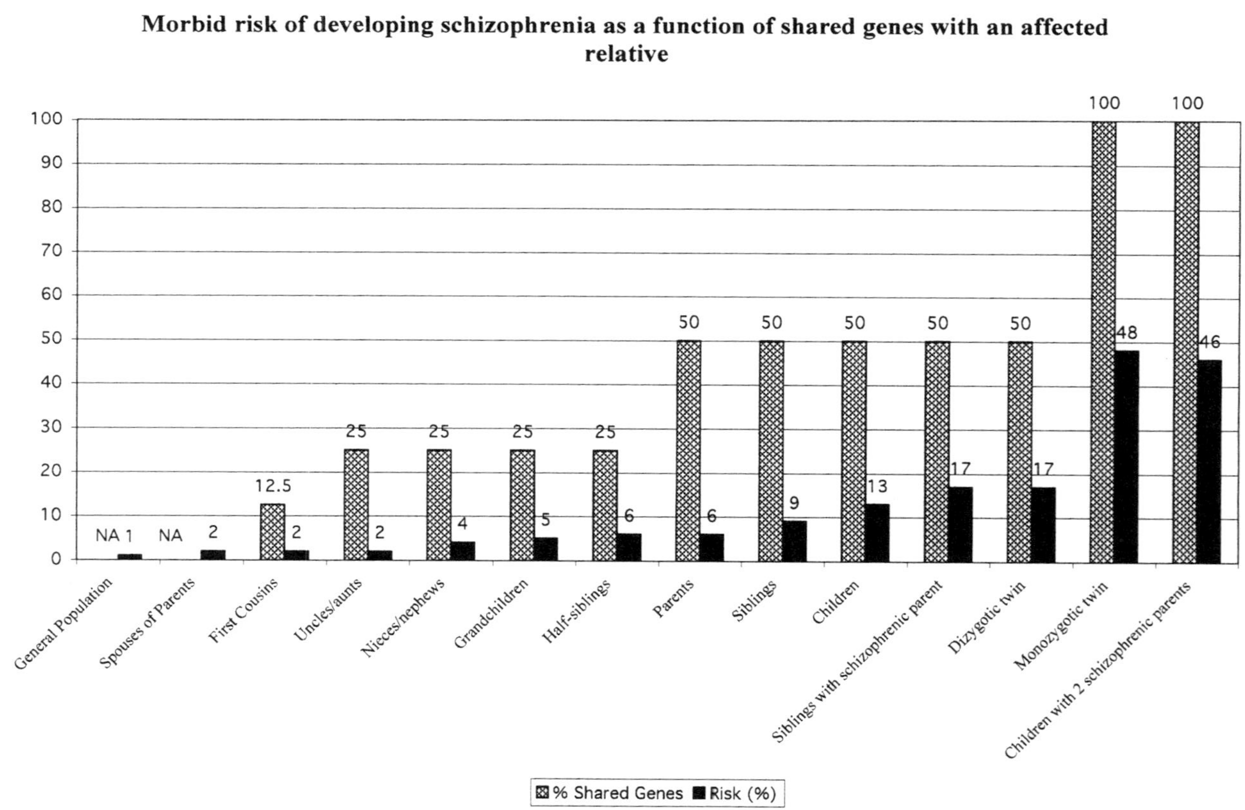

Figure 2 Graph showing risk for developing schizophrenia as a function of shared genes. Values are estimates, derived form Tsuang (Biol. Psych., 2001; originally complied by Gottesman, 1991)

	MZ		DZ		
	C	N	C	N	Reference
Adult disorders					
Schizophrenia	.48	115	.17	184	Gottesman 1991[b]
Affective illness	.65	146	.14	278	Berrettini 1997[b]
Alcoholism - Male	.41	413	.22	617	McGue 1995[b]
Alcoholism - Female	.34	155	.31	154	McGue 1995[b]
Criminal conviction	.52	229	.23	316	Gottesman & Goldsmith 1994[b]
Panic disorder	.24	67	.11	55	Kendler et al 1993a[c]
Bulimia nervosa	.23	35	.09	23	Kendler et al 1991[c]
Childhood disorders					
Attention deficit/Hyperactivity	.58	69	.31	32	Sherman et al 1997[c]
Tourette syndrome	.53	30	.08	13	Price et al 1985[c]
Autism	.64	45	.09	36	Smalley et al 1988[b]
Juvenile delinquency	.91	55	.73	30	Gottesman & Goldsmith 1994[b]
Reading disorder	.68	186	.38	138	DeFries & Alarcón 1996[c]

[a]When possible probandwise rather than pairwise concordance is reported.
[b]Concordance rate reflects the weighted average of studies reviewed in citation.
[c]Concordance rate from a single study, no compilation available.

Table 2 Monozygotic (MZ) and dizygotic (DZ) twin concordance (C) for behavioral disorders. Reproduced with permission from: McGue, M. & Bouchard, T. Genetic and Environmental Influences on Human Behavioral Differences. Annual Review of Neuroscience 1998; 21: 9.

The susceptibility and immune response to infectious agents is well known to be under genetic control. As discussed in Chapter 2, gene products can act at multiple levels to modify disease outcome following exposure to a virus. Heritability estimates for common childhood infections range from 5% (for mumps) to 86% (for rubella) (Gedda et al. 1984). Because twins and siblings tend to share the same physical space and personal contacts, viruses that are highly infectious generally have a high concordance rate in pairs of monozygotic and dizygotic twins. If one twin, for example, contracts measles from a schoolmate he has a high likelihood of spreading that virus to the other twin at home. Concordance rates for measles infection are similar for monozygotic and dizygotic twin pairs, 94% and 97%, respectively. On the other hand, for viral diseases that are less infectious or produce disease in only a subset of infected individuals the difference in concordance rates for monozygotic versus dizygotic twins tend to be greater. A case in point is paralytic poliomyelitis, which is an unusual complication of infection with poliovirus. The concordance rate for paralytic poliomyelitis in monozygotic twins is 36% compared to only 6% for dizygotic twins, indicating the importance of genetic factors in determining the outcome of exposure to this virus (Herdon et al. 1951). Along the same lines, rates of infection with hepatitis B virus are similar when comparing dizygotic or monozygotic twin pairs. But if hepatitis B carrier status is considered, for which the interactions between the virus and host genotype are presumably more important, then the higher concordance rates for monozygotic twin pairs is revealed (Lin et al. 1989).

In the strongest version of the viral hypothesis for schizophrenia, infection with a single (currently unidentified) virus is a necessary antecedent of all cases of schizophrenia. There is nothing in the current psychiatric genetics literature to refute this. Indeed, even if predisposing genes for schizophrenia are found be expressed exclusively in the CNS, this does not discount the viral hypothesis. It might simply imply that the virus and the gene product are acting at different points in the causal chain that eventually leads to schizophrenia symptoms. And while it's conceivable that a small subset of patients may have a purely genetic form of schizophrenia, the hopes of finding any one gene that underlies the disorder are diminishing. Instead, there is an emerging consensus that the genetic component to schizophrenia involves multiple interacting susceptibility genes (Jablensky 1999; Murray et al. 1992; Tsuang 2000). Some or all of these genes may be involved in moderating the outcome of infection with a virus.

Thus far, genome-wide screens hunting for specific genes responsible for schizophrenia have been disappointing and have yielded inconsistent results (DeLisi 2000). Even so, deciphering underlying genetic vulnerabilities represents one of the most promising areas in schizophrenia research, and candidate genes have included some that would be expected to play a role in susceptibilities or responses to viral infections. For example, recent studies have found indications that polymorphism in cytokine genes (IL-1 and TNF-alpha) may be associated with schizophrenia.

4.1.2 Season of Birth:

The season of birth effect is one of the most extensively studied phenomena in schizophrenia epidemiology. Since the first such study in 1929, there have been in over 250 published reports examining seasonal birth factors in patients with psychosis. A recent review scrutinized 86 studies of seasonal birth patterns among patients who developed schizophrenia, and controls (Torrey et al. 1997). Collectively, these studies represented 437,710 individuals from the northern hemisphere, and 25,826 from the southern hemisphere. The conclusion drawn from this comprehensive review was that there is a remarkably consistent increase of 5 to 8% more births occurring in winter-spring among patients who eventually develop schizophrenia compared to what would be expected by chance. This very well-replicated finding could not be explained by statistical artifacts or aberrations in control selection. The large number of patients examined in the northern hemisphere provided the most definitive picture of the phenomena, revealing that the peak excess of births in schizophrenia patients occurred during January and February. An example of a one season of birth study is shown in figure 3.

Considering that the birth-season factor in schizophrenia is bona fide and is now well accepted, a multitude of hypothesis have been put forth to explain the finding. While the majority of these are intriguing, and some are quite plausible (seasonal differences in maternal nutrition, procreation proclivities, photo-period and circadian effects, gene-temperature interactions, etc.), none have the face validity of an infectious etiology.

In this regard, one of the best-studied cyclical trends in epidemiology is seasonal variations in human infectious diseases. You need only to consider the common term "flu season" to realize that most of us are impacted by such cycles. But depending on the virus species, peak incidence may occur in any season. In terms of the neurodevelopmental hypothesis for schizophrenia, the long gestation of humans implies that the putative infection responsible for the seasonal effect is not necessarily acting at the time of birth. A winter birth may simply mark the end of the preceding gestational epochs that are obviously paralleled by the preceding seasons. It's possible that a winter-spring birth is a risk factor for premature birth-- which itself is a risk factor for schizophrenia. So far, however, there is not much support for the idea that winter births are simply surrogate markers of prematurity (Torrey et al. 1997). Thus a virus may well be responsible for the winter-spring seasonal birth increase, but the season during which the virus presumably acts cannot be discerned from these epidemiological studies.

Beyond the influence of seasons on the communicability of viruses, exposure to sunlight is well recognized to modulate the replication and pathogenesis of viruses. This extends to neurotropic viruses such as herpes simplex, which is prompted to reactivate by ultraviolet light exposure. Therefore the seasons are (in a sense) immunomodulators.

There are a couple of points that reduce the potential importance and cogency of a viral connection with the birth pattern seen in schizophrenia. First, the seasonal effect is rather small, accounting for only a fraction of schizophrenia cases. Granted, this could be because the seasonal variation of the underlying virus is also small, and hence, the virus might be present year round but may surge only slightly in the winter and spring. Such a virus could, in theory, be the cause of all cases of schizophrenia.

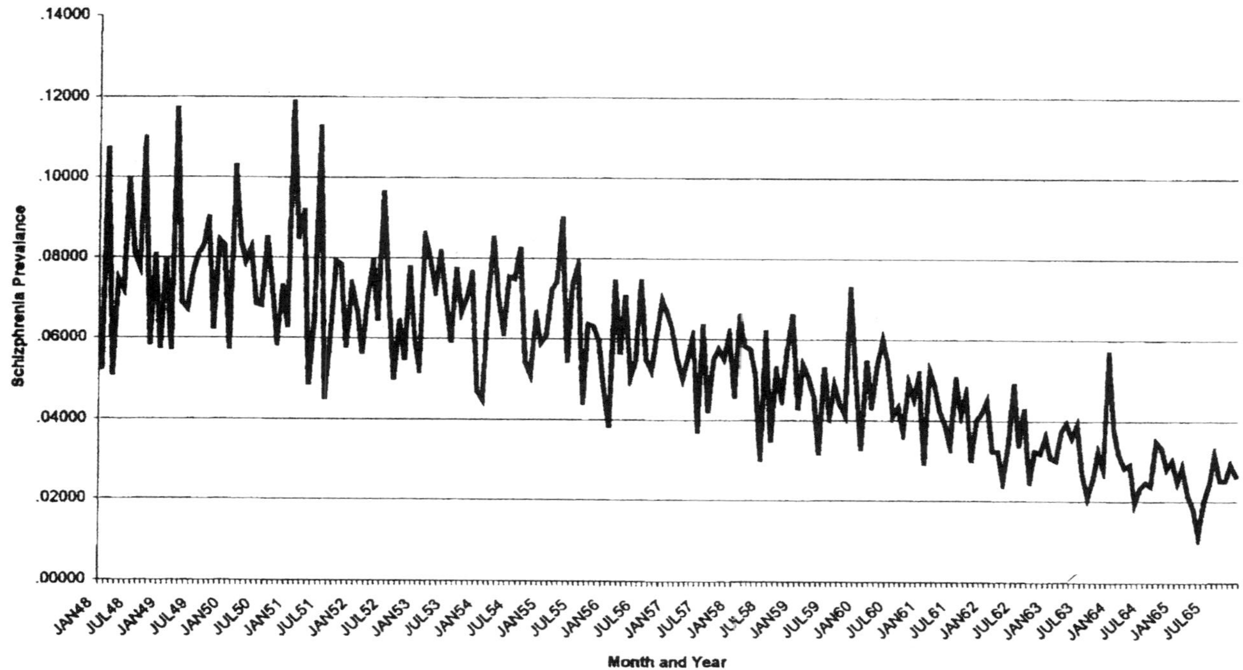

Figure 3 Monthly prevalence of schizophrenia in the state of Georgia from 1948 – 1965. The downward slope of the prevalence rates in the figure; is a result of artifact from using birthdates of the Medicaid recipients. By using participants' dates of birth as seasonal markers, the sample of individuals eligible for a schizophrenia diagnosis naturally decreases with each year. The age prevalence effect does not interfere with overall seasonal prevalences of schizophrenia. Republished with permission from Battle, et al., J. Psychiatric research 33:504)

The second problem with the season of birth effect is that it has been found for other neuropsychiatric disorders besides schizophrenia (Torrey et al. 1997). While some of these other conditions (bipolar disorder, autism, multiple sclerosis) have also been posited to be caused by a virus, this non-specificity of the seasonal effect dilutes the argument for a viral cause of schizophrenia.

4.1.3 Geographic Variation

When a disease shows clusters of cases in distinct geographical regions, or when the incidence changes abruptly over time, an infectious or environmental cause is most often suspected. Therefore it is not surprising that geographic patterns of schizophrenia have often been considered in the context of the viral hypothesis.

The worldwide prevalence of schizophrenia appears to be fairly uniform despite problems with detection, classification, and cultural influences on symptoms and diagnostic criteria (Jablensky 1999). In most countries surveyed the prevalence of schizophrenia is between 0.2% and 1%. Nevertheless there is evidence for pockets of high prevalence (e.g. northern Sweden, 1.7%) and low prevalence (Hutterites in Northwestern United States, 0.1%). There is also evidence for a temporal change in schizophrenia incidence. However, these increases or decreases in schizophrenia incidence do not resemble the typical pattern of an abrupt infectious disease "outbreak", but rather appear to be secular trends occurring over a more extended period of time. Secular trends are well-established for infectious illnesses but they are also characteristic of many (perhaps most) other diseases as well. When viewed broadly, regional or temporal patterns of schizophrenia occurrence cannot tell us much of the viral origin for the disease. Many viruses have a relatively even distribution across the globe. Conversely, genetic diseases can show distinct geographical clustering (e.g. due to inbreeding). A disease that involved the interaction of a virus with host susceptibility genes could have a distribution like schizophrenia-- but not necessarily.

Geographic and temporal trends for schizophrenia have been much more informative when subjected to more refined and detailed analysis. Studies by several different research groups have found that being born or raised in an urban environment is a risk factor for schizophrenia (McDonald et al. 2000; Mortensen et al. 1999). These epidemiological findings were recently crowned by a series of reports in which both family mental-health history and urbanization were measured (Mortensen et al. 1999; Pedersen et al. 2001). These study found that neither a family history of schizophrenia (genetic loading) or downward socio-economic drift could account for the urban birth increase. In fact, because so many people live in the city, the population attributable risk (the portion of people that would be theoretically spared the disease if the risk factor was removed) was greater for the city birth-factor than it was for having a parent with the disease. Thus, this environmental factor is of great importance to the prevalence of schizophrenia.

For many infectious diseases an increased incidence and younger age at infection is associated with urban dwelling. One of the factors responsible for this is household crowding (Torrey et al. 1998). The issue of whether being raised under crowded conditions increases one's risk for schizophrenia has not been resolved, but in any case there are other factors beyond household crowding that lead to higher rates of viral infections in an urban setting.

For example, schizophrenia could be caused by a virus transmitted from house cats (Torrey et al. 1995). This hypothesis is consistent with the urban excess of births because cats tend to be indoor pets in the city whereas they are more often kept outside in rural areas. Perinatal exposure to cats could also explain the increase of pre-schizophrenia births during the winter and spring months when owners bring their cats into the house to protect them from the cold and rain. Cats are known to harbor a number of viruses, including retroviruses, and transmission of a human neuroteratogenic protozoan (Toxoplasma gondii) from house cats to their pregnant owners is well established. Interestingly, a recent report measuring antibodies against Toxoplasma gondii in first episode schizophrenia patients found quantitatively higher amounts of IgG and IgM in the patients compared to controls (Yolken et al. 2001). In general, this protozoan is not thought to invade the CNS of adult humans with normal immune systems, but there could have been unidentified immunological variables in schizophrenia patients that caused them to produce these antibodies.

Some viruses are more common in rural areas--e.g. because of more primitive waste disposal or more contact with farm animals. In theory, exposure to viruses encountered in the countryside could be protective against subsequent infection with a "schizophrenogenic" virus. In addition, socioeconomic status has a profound impact on a person's infectious disease history, and lower socioeconomic status is clearly a risk factor for schizophrenia, (though the degree to which schizophrenia causes, or is caused by, social and economic variables has not been untangled).

4.1.4 Migration and Risk.

The study of schizophrenia rates in migrant populations also supports the viral hypothesis. History has firmly established that immigrants easily fall victim to viruses that are not endemic in their country of origin. Because the immigrants lack the "herd immunity " present in the populace of their host country, a wave of infection may sweep through the new arrivals. The shape of this wave may be sharp or more gradual depending on the immigration pattern and the virus transmission-dynamics.

Numerous studies have reported increased rates of schizophrenia associated with immigration to a new country (McDonald & Murray 2000). This phenomenon has been reported for inter- and intra- continental migrations, but the excess of schizophrenia among Afro-Caribbean migrants to England has been most extensively studied.

From the mid 1950's until 1960 there was a large wave of migration from the Caribbean to England. Early reports suggested that schizophrenia appeared to be higher among Afro-Caribbean people living in England compared to the white population. The implication of these early findings are unclear because of the issues related to misdiagnosis and classification, and because it was obscure how many of the Afro-Caribbean patients were born in the Caribbean versus England.

Subsequent studies, focusing on the first generation offspring of these Afro-Caribbean migrants, were much more revealing (McGovern et al. 1987; Sharpley et al. 2001). These studies found that children who were born in England of parents who had migrated from the Caribbean to England were much more likely to develop schizophrenia than either their counterparts still living in the Caribbean, or white people born in England. Additional studies suggested this phenomenon could not be ascribed to misdiagnosis. Most importantly, genetic factors were not responsible because the enhanced risk of being a first generation (England-born) person of African-Caribbean migrant parents was not reflected as enhanced risk in the migrant parent. Thus, there is no evidence of excessive genetic loading in this first generation of (England-born) Afro-Caribbeans who were shown to be at a substantially elevated (approximately 15 fold) risk of developing schizophrenia. Accordingly most of the increased risk has been attributed to environmental factors, and viral infections remain one of the prime candidates.

Several scenarios in which a virus could be the culprit in migrant populations can be considered. Since the Afro-Caribbean patients were British-born, the increased risk could result from an enhanced susceptibility to a virus on the part of the migrant mother. There is a precedent for this mechanism in that Afro-Caribbeans migrating to England in the 1950's were particularly susceptible to rubella (Sharpley et al. 2001). Interestingly, a recent report has implicated rubella as one of the developmental insults leading to schizophrenia; and congenital rubella has indeed been found to be higher among the offspring of young Afro-Caribbean women who contracted the virus after their migration to England (Brown et al. 2000; Sharpley et al. 2001). Seemingly, this could help account for that first generation effect discussed above, but this cohort of schizophrenia patients actually experienced fewer severe pregnancy complications than their white counterparts. Nevertheless, a congenital infection that was less severe and less noticeable than rubella could certainly be responsible.

Infection occurring during infancy or early childhood has also been proposed to cause schizophrenia. It's possible that the first generation of British-born Afro-Caribbeans contracted the virus postnatally. The lack of maternal immunity could permit infection of the neonates at an inopportune time in their neurodevelopment whereas most white people would be protected by maternal immunity to the endemic virus. The finding of a higher rate of schizophrenia in the siblings of affected first generation patients is consistent with the idea that postnatal infection of the index case would also occur in his sibling because both would be lacking protective maternal antibodies.

The large effect-size in this Afro-Caribbean cohort could be a consequence of "two hits"; the first hit being an infection by a virus during neurodevelopment, and the second hit being psychosocial stressors arising from attempts to cope with economic adversity and racism as the patient matures. In such a scenario, the unusually high rate of schizophrenia in this population could be derived from an especially high "penetrance" of genetic and viral factors brought to the surface by psychosocial triggers.

It is unlikely that the viral hypothesis of schizophrenia will be proved or disproved based solely on geographic patterns of disease. Nevertheless, such studies can lend support to existing sub-theories, or they can generate their own new sub-theories. An interesting example of the latter is a study by Brown who found that states with the highest rates of schizophrenia also had the greatest population of ticks capable of spreading tick-born encephalitis viruses (Brown 1994). The author hypothesized that schizophrenia may be caused by a vector-borne flavivirus, contracted either during gestation or later in life.

There are numerous species of flaviviruses throughout the world, but these often have distinct and non-overlapping ranges. Moreover, tick-borne diseases are usually thought to be associated with rural environments, a notion that contradicts the geographic distribution of schizophrenia. But as correctly pointed-out by Brown, tick-borne infections in urban centers may be under-appreciated. Brown's hypothesis would be strengthened by the discovery that most flaviviruses share some common pathogenic mechanisms that could link them to schizophrenia (all tend to be in neurotropic, but most probably do not cross the placenta). It would also be supported if there were a higher rate of seropositivity for flaviviruses in patients with schizophrenia. The current serological data are incomplete but are not particularly supportive. However, as noted by Rubinstein, the critical factor linking flaviviruses with schizophrenia may be the presence of a resistance gene against this set of pathogens (Rubinstein 1997). She postulates that individuals who possess a certain allele that confers resistance to flaviviruses have, in a sense, made a trade-off because that gene may also be associated with hypersensitivity to dopamine. While some microbial resistance genes obviously exist in geographically distinct patterns, being more frequent where the disease is endemic (e.g. resistance to malaria), what is not clear in the flaviviruses example is why the proposed flaviviruses resistance gene would be expected to "breed true" across state lines as implied by Brown's data. Regardless of the caveats, because resistance genes for flaviviruses and Lyme arthritis (caused by a tick-borne bacterium) may be linked on the same chromosomal segment, and since arthritis is very low in schizophrenia, this hypothesis has heuristic value and is continuing to be investigated. Indeed, Brown has recently proposed that the flavivirus resistance gene may be altering protein translation, including proteins found to be abnormal in schizophrenia.

4.1.5 Gender

Overall, schizophrenia affects males and females in approximately the same proportions. However, on the average males are younger at disease onset and have a more severe course. There are also sex differences for the vast majority of infectious diseases. Depending on the organism these can involve a multitude of differences including: the age at first infection, incidence rates, disease severity, and characteristics of sequela. There may well be important pathophysiological relationships hiding among the gender differences in both schizophrenia and viral diseases that could provide important clues to the viral hypothesis. Perhaps the reason there have been few attempts to investigate these relationships is because the interwoven social, psychological and physiological determinants of sex differences governing the acquisition and response to infectious diseases is no less complex than those governing the development of schizophrenia.

4.2 SEROEPIDEMIOLOGICAL STUDIES

Seroepidemiology is the study of infectious disease patterns in groups or populations by examining antibody responses in the serum or plasma. The most common statistic in seroepidemiology is seroprevelenace, which is defined as the portion of the population that demonstrates the presence of an antibody to the organism of inquiry. In reality, seroprevalence is not a measure of disease prevalence but is a measure of partial cumulative incidence because many people with antibodies to a virus have cleared the organism and are not suffering from an infectious disease, and other people tested by serology may have been infected with the virus but never made an adequate antibody response, or had serum antibodies that declined to undetectable levels during the interval between infection and testing.

The most striking aspect of serological studies in schizophrenia is the sheer volume of work that has been performed. This can be gleaned from tables 2 and 3 showing some of the findings for herpes simplex as well as an assortment of other viruses. Indeed, since many infectious disease specialists are not familiar with the viral hypothesis for schizophrenia they are (in my experience) often quite surprised to learn how frequently this diagnostic and investigative serological technique has been applied to a psychiatric disorder.

The rationale for investigating a particular virus in these studies may be built-up from the known pathophysiological properties of an organism, e.g. a neurotropic herpes viruses, or it may stem from the motivation to "look where the light is"; that is, an assay to detect antibodies against a certain virus becomes available and is consequently applied to schizophrenia serum with the hopes of finding a positive result.

The herpes viruses have been most extensively examined, and there are some well-performed studies showing increased seroprevalence for HSV and CMV in schizophrenia. But there are also numerous negative studies. On the whole, the most common finding is that there are no significant differences in seropositivity when comparing schizophrenia patients with various control groups. If we isolate those studies which did find a difference, an increased incidence of seropositivity or antibody titer in schizophrenia is a much more common finding than reports of a decrease. While this could be due to publication bias, a decrease would not necessarily be uninteresting, and the fact that many of the articles reported results for several different viruses mitigate this interpretation.

Table 3. Antibodies to Herpes Simplex Virus in Schizophrenia Serum.

Case vs. Control	Body Fluid	Number of SCH	Control Group	Comments	Author * (Year)
_	Serum	38 medicated SCH	22 NPC matched on age, sex, and race	No significant difference (p=0.23)	Pelonero (1990)
_	Serum	12 non-medicated SCH	10 NPC	No significant difference	Rimon (1978)
_	Serum	16 severely psychotic SCH	18 surgical NPC	No increased titers in SCH	Rimon (1979)
_	Serum	9 schizo-affective, 43 SCH	10 surgical NPC (no known mental disorders and 80 neurological disease	No increased titers in SCH	Torrey (1978)
_	Serum	38 SCH	17 with no history of major psychiatric disorder in 1st-degree relative; and 24 hospital staff	No significant difference	DeLisi (1986)
_	Serum	3 non-medicated and 27 medicated SCH	30 NPC	No significant difference	Schindler (1986)

Case vs. Control	Body Fluid	Number of SCH	Control Group	Comments	Author* (Year)
_	Serum	48 SCH	48 patients with other psychosis and 48 hospital staff	No significant differences	Fux (1992)
_	Serum	11 SCH	9 age and sex matched hospital staff	No significant differences	Fukuda (1999)
_	Serum	61 SCH	34 hospital staff	No significant difference	Rimon (1969)
_	Serum	68 SCH	38 NPC	No significant difference	Pokorny (1973)
_	Serum	327 SCH	387 NPC and 100 blood donors	No significant difference	Lycke (1974)
↑	Serum	54 SCH	32 hospital staff from same hospital as patients	Antibody higher in SCH ($p<0.001$)	Halonen (1974)
_	Serum	41 SCH	25 patients with neurologica l diseases	No significant difference	Gotlieb-Stematsky (1981)
↑	Serum	265 SCH:	33 hospital staff and 387 NPC matched on residence	Neutralizing antibody titers were higher in schizophrenics vs. controls ($p=0.001$)	Libikova (1983)

Case vs. Control	Body Fluid	Number of SCH	Control Group	Comments	Author* (Year)
_	Serum	20 SCH	17 surgical NPC matched on gender and age	No significant difference	King (1985b)
_	Serum	127 SCH	52 NPC	No significant difference	Cazzullo (1987)
_	Serum	7 SCH	21 NPC	No significant difference	Gotlieb-Stematsky (1987)
↑	Serum	233 SCH	215 NPC matched on age	No significant difference in ELISA; mean neutralizing Ab titers higher ($p=0.001$)	Rajcani (1991)
_	Serum	222 SCH	30 hospital staff, 76 blood donors, and 37 elderly NPC	No significant difference	King (1985a)
_	CSF	12 non-medicated SCH	10 NPC	No significant difference	Rimon (1978)
_	CSF	16 severely psychotic SCH	18 surgical NPC	No increased titers in SCH	Rimon (1979)

Case vs. Control	Body Fluid	Number of SCH	Control Group	Comments	Author* (Year)
–	CSF	9 schizo-affective, 43 SCH	10 surgical NPC (no known mental disorders and 80 neurological disease	No increased titers in SCH	Torrey (1978)
–	CSF	20 SCH	17 surgical NPC matched on gender and age	No significant difference	King (1985b)

* If more than one author, only first author is indicated.

Serum indicates serum or plasma.

[1]↑ = increase

↓ = decrease

– = no difference

[2]SCH = Schizophrenia Cases

[3]NPC = Non-psychiatric Controls

Table 4. Serology for Viral Antibodies in Schizophrenia

Virus[1]	Studies Showing Increase	Studies Showing No Change	Studies Showing Decrease	References[2]
Adenovirus	0	**3** (2 serum, 1 CSF)	0	17, 18
Arbovirus	0	**1** (serum)	0	19
BDV	**4** (all serum)	**3** (all serum)	0	3, 7, 13 – 15, 25, 30
CMV	**3** (all CSF)	**16** (12 serum, 4 CSF)	0	2, 5, 6, 8, 16 – 21, 23, 24, 26 – 29
EBV	**3** (all serum)	**4** (all serum)	0	2, 5, 6, 9, 10, 17, 21
HIV – 1	**1** (serum)	0	0	12
HTLV – I	0	**1** (serum)	0	4
HTLV – III	0	**1** (serum)	0	4
HHV – 6	0	**1** (serum)	0	6
Influenza A	**1** (CSF)	**4** (3 serum, 1 CSF)	0	1, 6, 29
Influenza B	0	**1** (serum)	0	6
Japanese Encephalitis Viruses	0	**2** (all serum)	0	6
Measles	**2** (1 serum, 1 CSF)	**10** (8 serum, 2 CSF)	0	6, 11, 17 – 22, 27

Virus[1]	Studies Showing Increase	Studies Showing No Change	Studies Showing Decrease	References[2]
Mumps	0	3 (all serum)	2 (1 serum, 1 CSF)	6, 17, 18, 21
Rubella	0	9 (6 serum, 3 CSF)	0	6, 11, 17, 18, 22, 27
TEV	1 (serum)	0	0	19
Vaccinia	3 (1 serum, 2 CSF)	0	0	1, 29
VZV	0	7 (6 serum, 1 CSF)	0	2, 6, 17 – 20

*Only papers in English were included. There are numerous studies for HIV in schizophrenia that are not included because a control group was not defined.
Serum indicates serum or plasma.

[1] BDV = Borna Disease Virus; CMV = Cytomegalovirus; EBV = Epstein-Barr Virus; HIV = Human Immunodeficiency Virus; HTLV = Human T-Cell Leukemia Virus; HHV = Human Herpesvirus; TEV = Tick-borne Encephalitis Virus; VZV = Varicella – Zoster Virus

[2] 1. Albrecht, P., E. F. Torrey, et al. (1980). 2. Cazzullo, C., C. Domonico, et.al. (1987). 3. Chen, C. H., Y. L. Chiu, et al. (1999). 4. De Lisi, L. E. and P. S. Sarin (1985). 5. De Lisi, L. E. (1986). 6. Fukuda, R., T. Sasaki, et al. (1999). 7. Fukuda, K., K. Takahashi, et al. (2001). 8. Fux, M. (1992). 9. Gotlieb-Stematsky, T., J. Zonis, et al. (1981). 10. Gotlieb-Stematsky, T. (1987). 11. Halonen, P. E., R. Rimon, et al. (1974). 12. Hart, D. J. (1999). 13. Horimoto, T. and H. Takahashi (1997). 14. Iwahashi, K. (1997). 15. Iwahashi, K., M. Watanabe, et al. (1998). 16. Kaufmann, C. A., D. R. Weinberger, et al. (1983). 17. King, D. J. (1985). 18. King, D. J. (1985). 19. Libikova, H. (1983). 20. Lycke, E., R. Norrby, et al. (1974). 21. Pelonero, A. L. (1990). 22. Rimon, R. (1978). 23. Rimon, R. (1986). 24. Schindler, L. (1986). 25. Selten, J. P., D. Wiersma, et al. (2000). 26a. Shrikhande, S. (1985). 27. Torrey, E. F. (1978). 28. Torrey, E. F. (1982). 29. Torrey, E. F., R. H. Yolken, et al. (1983). 30. Yamaguchi, K. (1999).

The only fair interpretation of the available data is that seroepidemiological studies have failed to pinpoint a particular virus underlying schizophrenia pathophysiology, and that this approach does not appear to be very efficient. This is not to say that this large body of data should be ignored, but merely that the discrepancies and inconsistencies between studies are reason for concern, and there is a need to integrate the data and formulate a plan for the future direction of this line of inquiry. The first step in doing this is to delineate possible interpretations of the available studies, some of which have indicated increased seropositivity among schizophrenia patients:

i) Higher rates of seropositivity among schizophrenia patients may be a consequence of increased exposure to certain viruses because of behavioral differences arising after the patient develops mental illness. Most clinicians could easily name a litany of lifestyle differences associated with schizophrenia that could increase contact with viruses or enhanced susceptibility to infection. These include: tobacco use, poor hygiene, unusual food sources, antipsychotic medication, and nosocomial exposures.

ii) It may well be that none of virus tested for by seroepidemiology have any etiopathological role in schizophrenia. The increased titers or rate of seropositivity found in some studies could, however, reflect a proclivity among schizophrenia patients to make a more vigorous antibody response when challenged with infection by some common viruses such as HSV or CMV. In such a scenario, schizophrenia may have an immunogenetic component that leads to aberrant antibody responses to viral infection, which in turn could play a role in the developmental or behavioral pathophysiology of the disease.

iii) One or more of the viruses listed in the tables could participate in the causal chain leading to schizophrenia. There is no reason to believe that a virus will produce detectable antibodies in all infected individuals. There is considerable inter-individual variability in how robust and sustained an antibody response may be following challenge with a viral infection. The repertoire of viral epitopes recognized by particular individuals may also influence the sensitivity of the immunoassay employed to evaluate seropositivity. The issue of detection and response magnitude is particularly pertinent to perinatal infections because fetuses and neonates often do not mount vigorous immune responses. Furthermore, even if a response is made, it could easily wane to undetectable levels in the decades between infection and antibody testing, which is performed after schizophrenia has developed. On the other hand, serological assays for viral antibodies can give false positives because of antibodies generated against related viruses, or even as a consequence of polyclonal stimulation arising from bacterial infection or autoimmune disease. Thus theoretically, an unidentified herpes virus could be causing all cases of schizophrenia; the infection could be manifested as an antibody response that cross reacts with CMV in only some immunoassays, or in serum only from those individuals whose antibody repertoire included binding to epitopes that were shared between the novel (unidentified) herpes virus and CMV.

Overall, the hunt for specific serological indices of infection in schizophrenia, like the hunt for specific genes underlying the disease, has been discouraging but has not reached the point of abandonment. Some headway may be made with the employment of new techniques, such as epitope peptide arrays, or, as new assays are developed based on candidate viruses identified through other lines of research.

4.3 IMMUNE ALTERATIONS IN SCHIZOPHRENIA

While the seroepidemiological studies discussed in the previous section may be of dubious of fidelity, studies of the immune system in schizophrenia leave no doubt that the disease is associated with anomalies affecting multiple levels of the immune axis (Ganguli et al. 1993). This being said, the information on immune dysfunction in schizophrenia is fragmentary and is rife with confounds. Like most investigations of schizophrenia, immunological studies are frustrated by the clinical heterogeneity of the disease, and the difficulty in determining whether a positive finding has an intrinsic causal role or is merely an epiphenomenon. The rationale for a causal role of immunological changes in schizophrenia pathophysiology rests squarely on a massive foundation of literature showing that immune cells and molecules can profoundly influence complex behaviors in humans and experimental animals.

The scientific literature on immunological alterations in schizophrenia is voluminous and fascinating. Studies of abnormal immune indices in schizophrenia date back more than a century, and in just the last 30 years there have been over 1000 papers dealing with some aspect of the subject. Hence, a thorough review of this area is beyond the scope of this book. Therefore the present short section reviews only a fraction of this literature as it pertains most directly to the viral hypothesis.

Viewed from the vantage point of the viral hypothesis of schizophrenia, the immune anomalies in the disease could be determinants of increased susceptibility to viral infections, or conversely, they could be manifestations of an abnormal host response leading to immune-mediated neuropsychiatric sequela. The types of immune studies in schizophrenia can be divided into three categories: studies of immune cell populations; measurements of cytokines and soluble inflammatory mediators; and investigations of autoimmunity (e.g. autoantibodies)(Ganguli et al. 1993; Kirch 1993; Muller et al. 1999).

For many viral and immune mediated disorders, the analysis of peripheral blood cell markers to characterize changes in cell populations mediating innate and acquired immunity has been well characterized. This has not been the case for schizophrenia. In schizophrenia the relative and absolute numbers of lymphocytes, NK cells, B cells, T-cells, and CD4/CD8 T-cell subpopulations have been variously reported to be increased, decreased, or unchanged. With the possible exceptions, the data concerning immune cell populations in schizophrenia is equivocal, and has yielded little insight into the viral hypothesis.

Studies of cytokines and inflammatory mediators in schizophrenia have been more illuminating, though the data are far from universal. Most studies have found an increase in interleukin-2 soluble receptor (IL-2sR) in schizophrenia, which is an indicator that T-cells are activated (Rapaport et al. 1994; Rapaport et al. 1993). The in vitro secretion of interleukin-2 itself appears to be decreased in schizophrenia, according to the majority (11 out of 18) of studies. This finding seems at odds with the increased IL-2sR finding that indicates immune activation. However immunological studies in other diseases have shown that increased IL-2sR and dampened in vitro production of stimulated IL-2 are consistent with a background of autoimmunity. Schizophrenia also appears to be associated with an increase in the proinflammatory cytokines IL-1 and IL-6, at least in a subset of patients. Also consistent with an autoimmune disease, the in vitro production of interferon gamma (IFN-gamma) may be decreased in lymphocytes from schizophrenia patients. Since CD4 lymphocytes are a major source of cytokines such as IL-2 and IFN-gamma, the question has been raised whether schizophrenia is associated with an imbalance between the two types of CD4 cells--the TH1 cells which produce IFN-gamma , IL-2 and IL-12, and the TH2 cells which produce IL-4, IL-5 and IL-10 (Schwarz et al. 2001). Although it is somewhat of an oversimplification, TH1 cells tend to promote cell-mediated immune responses whereas TH2 cells aid in the maturation of B cells and encourage antibody responses. Because TH2 responses act to suppress TH1 activity (and visa versa), the decreased production of some TH1 cytokines in schizophrenia suggests that TH2 activity may be taking ascendancy over TH1 responses. This could encourage autoantibody formation--but this is largely a matter of speculation. Still, the preponderance of the data indicate that schizophrenia is associated with a dysregulation of CD4 helper cells, and that the immune anomalies in schizophrenia may have a genetic basis that predisposes these patients toward certain types of autoimmune responses. In this regard, an autoimmune basis for schizophrenia has been considered by numerous investigators.

One of the more straightforward of such propositions is that schizophrenia is caused by circulating antibodies present in the plasma and/or CSF that cross react with brain proteins. Proving this hypothesis, however, has not been straightforward and has generally been met with skepticism. Some of the attempts to find these cross-reacting antibodies are listed in table 5.

Table 5

Brain Structure/ molecule	SCH vs. Control[1]	Body Fluid	Control Group	Author (Year)
Brain Septal Region	↑	Serum	117 NPC[2]	Baron (1977)
IMR-32 Human Neuro-blastoma Cell Line	↑	Serum	22 hospital staff	Mazeh (1998)
Septal Region	↑	Serum	72 NPC	Heath (1989)
Amygdala, Frontal Cortex, Cingulate Gyrus, Septal Area	↑	Serum	50 NPC age and sex matched	Henneberg (1994)
Human Brain S100 Protein, Neuron Specific Enolase (NSE), Myelin Basic Proten (MBP)	↑	Serum	112 blood donors	Jankovic (1991)
Cerebrum	↑	Serum	42 NPC	Sugiura (1989)
Fronto-cortical, Limbic, Extra-pyramidal Antibodies	↑	Serum	27 NPC	Pandey (1981)
Anticerebral Antibodies	↑	Serum	54 NPC	Shima (1991)
Fronto-cortical, Limbic, Extra-pyramidal Antibodies	↑	CSF	27 NPC	Pandey (1981)
Neural Antibodies	↑	Serum	58 NPC	Kilidireas (1992)
Septal Region, Caudate Nucleus	↑	Serum	19 NPC	Heath (1967)
Ganglioside	↑	Serum	31 NPC and 20 neuro-logical patients	Stevens (1992)
Nicotinic Acetylin-choline Receptor (nAChR)	↑	Serum	10 NPC	Mukherjee (1994)
Serotonin	↑	Serum	10 NPC sex and age matched	Levy-Soussan (1996)
Dopamine	↑	Serum	10 NPC sex and age matched	Levy-Soussan (1996)

Refer to Henneberg (1994) for more autoantibody studies.
Serum indicates serum or plasma.
[1]↑ = increased or more detected in schizophrenia
[2]NPC = Non-psychiatric controls, SCH=Schizophrenia

This table shows some of the numerous positive studies, but there have been a comparable number of negative studies, many examining the same brain regions.

As alluded to in Chapter 2, identifying cross-reactive epitopes in an autoimmune disease can be a difficult task because the primary and tertiary structure of the antigens are usually not known. In fact, for many established humoral autoimmune diseases of humans the antigen-antibody pair responsible for the pathological alterations has not been identified. To further complicate matters, autoimmune diseases do not just involve antibodies but are a consequence of complex interactions between predisposing genes and environmental factors that are played out according to a dynamic network of immunoregulatory cells and molecules.

There is a solid precedence for autoimmune diseases inducing psychotic states. The classic autoimmune disease, systemic lupus erythematosus, has been reported to give rise to psychosis in as many as one-quarter of the cases. Autoantibodies produced during some lung cancers can cause limbic encephalitis--with associated neuronal loss in the hippocampus and amygdala, and symptoms which include seizures and hallucinations (Arnason 1998). One of the more interesting psychiatric syndromes mediated by autoimmunity is Sydenham's chorea, which arises following infection with the beta-hemolytic streptococcus (Swedo et al. 1997). This organism is notorious for generating autoantibodies, and in this disease the humoral attack appears to be directed against neurons in the basal ganglia. Besides abnormal movement some patients demonstrate emotional changes, irrational fears, and obsessive compulsive behavior. Most patients recover, but Sydenham's chorea may represent only one disorder in a spectrum of pediatric autoimmune neuropsychiatric diseases that are triggered by infection with streptococcus.

If schizophrenia is an autoimmune disease, then we might expect patients with schizophrenia to have a higher rate of other autoimmune diseases. In this respect, it might seem surprising that one of best-replicated findings in schizophrenia epidemiology is the extremely low rates of rheumatoid arthritis (a classic autoimmune disease) among schizophrenia patients (Torrey et al. 2001). This resistance to rheumatoid arthritis is not an antipsychotic medication effect as it was first reported in 1936. On closer inspection, however, the arthritis connection could be consistent with an autoimmune basis for schizophrenia. Rheumatoid arthritis is a disease that has a genetic component, mediated at least in part by MHC alleles, and is triggered by an environmental insult--most commonly believed to be an infection. Somewhat similarly, schizophrenia has a genetic and environmental component and has been tentatively associated with different MHC genes than those thought to be involved in rheumatoid arthritis. In humans the MHC gene complex resides on the short arm of chromosome 6--

one of the hot spots in gene linkage studies in schizophrenia. Besides gene products responsible for antigen presentation (MHC-1 and MHC II proteins), the MHC gene complex (usually called the HLA complex in humans) contains a multitude of genes controlling various aspects of the immune response.

Thus it's possible that depending on which set of genes an individual has inherited at the MHC complex, a challenge with a virus will lead the immune system down one or the other of two alternative paths. In this simplified scenario, one path would lead to an immune cascade that had a relatively high likelihood of engendering immune conditions conducive to the development of rheumatoid arthritis; conversely, if the individual's genetic endowment dictated the alternative path, then the network of cytokines and immune mediators would direct the antiviral immune response in a different direction-- a direction that would allow for autoantibodies to be made against CNS components. Evolution necessitates that immune diversity be maintained in a species in order to prevent pathogens from mutating to avoid immune surveillance. On an individual level, the host has made a genetically preprogrammed trade-off between two potentially deleterious consequences of mounting an antiviral response.

This concept of alternative immune responses to a common agent is implicit in several sub-theories attempting to grapple with the role of the immune system in schizophrenia. It is also congruent with the suggestion that exposure to house cats is a feature that is shared in common between patients with schizophrenia in those with rheumatoid arthritis (Torrey & Yolken 2001). Specifically, as alluded to above, the cat could be a vector for an organism that pressed the immune system to invoke one of these two alternative paths described above.

A purist might argue that there is no definitive example of a clearly-identified virus causing autoimmune disease in humans. However, considering the plethora of evidence derived from studies in experimentally-infected animals, and in humans with autoimmune diseases, as well as the known ability of bacterial pathogens to trigger autoimmunity in humans, most researchers in this area have now turned their attention to defining the mechanisms involved (Moller 1998; Oldstone 1989; Strassburg et al. 1996). There are several mechanisms by which a viral infection may break immune tolerance and permit the immune response to be directed against self. In a process known as molecular mimicry, a protein or polysaccharide on the virus may be structurally homologous to a host molecule. The immune system, being unable to distinguish the two, may then cross-react with host cells and tissues expressing this molecule. In another mechanism, the virus causes the release into the circulation of autoantigens that are normally hidden from the immune system. Another way in which tolerance may be broken, at least theoretically, is due to the ability of the viruses to pick up host proteins as it buds from the cell membrane. These host-cell derived proteins may then become immunogenetic since they are present on the virus particle. In an analogous mechanism, the virus in the process of replication, may structurally alter host proteins that then become recognized as foreign to the immune system. There is substantial work to be done to prove the relevance of each of these mechanisms in autoimmunity in general, and in schizophrenia in particular.

In teleological terms, viruses are one of the main reasons that the immune system exists. Hence it is not surprising that viral and immune theories of schizophrenia are closely intertwined. By recognizing this, both viral and immune hypotheses stand a better chance of hitting their mark.

REFERENCES

Arnason BGW.Autoimmune diseases of the central and peripheral nervous systems. In: NR Rose; IR Mackay, editors The Autoimmune Diseases. San Diego: Academic Press; 1998; p. 571-602.

Beasley RP. Hepatitis B virus. The major etiology of hepatocellular carcinoma. Cancer 1988; 61:1942-56.

Brown AS, Cohen P, Greenwald S et al. Nonaffective psychosis after prenatal exposure to rubella. American Journal of Psychiatry 2000; 157:438-43.

Brown JS, Jr. Geographic correlation of schizophrenia to ticks and tick-borne encephalitis. Schizophrenia Bulletin 1994; 20:755-75.

Cardno AG, Marshall EJ, Coid B et al. Heritability estimates for psychotic disorders: the Maudsley twin psychosis series. Archives of General Psychiatry 1999; 56:162-8.

DeLisi LE. Critical overview of current approaches to genetic mechanisms in schizophrenia research. Brain Research Reviews 2000; 31:187-92.

Ganguli R, Brar JS, Chengappa KNR et al. Autoimmunity in schizophrenia: a review of recent findings. Ann. Med. 1993; 25:489-96.

Gedda L, Rajani G, Brenci G et al. Heredity and infectious diseases: a twin study. Acta Geneticae Medicae et Gemellologiae 1984; 33:497-500.

Herdon CN, Jennings RG. A twin-family study of susceptibility to poliomyelitis. Human Genetics 1951; 3:17-46.

Jablensky A. Schizophrenia: Epidemiology. Current Opinion in Psychiatry 1999; 12:19-28.

Kirch DG. Infection and autoimmunity as etiologic factors in schizophrenia: a review and reappraisal. Schizophrenia Bulletin 1993; 19:355-70.

Lin TM, Chen CJ, Wu MM et al. Hepatitis B virus markers in Chinese twins. Anticancer Research 1989; 9:737-41.

McDonald C, Murray RM. Early and late environmental risk factors for schizophrenia. Brain Research - Brain Research Reviews 2000; 31:130-7.

McGovern D, Cope RV. First psychiatric admission rates of first and second generation Afro Caribbeans. Social Psychiatry 1987; 22:139-49.

McGue M, Bouchard TJ, Jr. Genetic and environmental influences on human behavioral differences. Annual Review of Neuroscience 1998; 21:1-24.

Moller E. Mechanisms for induction of autoimmunity in humans. Acta Paediatrica. Supplement 1998; 424:16-20.

Mortensen PB, Pedersen CB, Westergaard T et al. Effects of family history and place and season of birth on the risk of schizophrenia. [see comments]. New England Journal of Medicine 1999; 340:603-8.

Muller N, Riedel M, Ackenheil M et al. The role of immune function in schizophrenia: an overview. European Archives of Psychiatry & Clinical Neuroscience 1999; 249 Suppl 4:62-8.

Murray RM, Jones P, O'Callaghan E et al. Genes, viruses and neurodevelopmental schizophrenia. Journal of Psychiatric Research 1992; 26:225-35.

Oldstone MB. Virus-induced autoimmunity: molecular mimicry as a route to autoimmune disease. Journal of Autoimmunity 1989; 2 Suppl:187-94.

Pedersen CB, Mortensen PB. Family history, place and season of birth as risk factors for schizophrenia in Denmark: a replication and reanalysis. British Journal of Psychiatry 2001; 179:46-52.

Rapaport MH, Lohr JB. Serum-soluble interleukin-2 receptors in neuroleptic-naive schizophrenic subjects and in medicated schizophrenic subjects with and without tardive dyskinesia. Acta Psychiatrica Scandinavica 1994; 90:311-5.

Rapaport MH, Torrey EF, McAllister CG et al. Increased serum soluble interleukin-2 receptors in schizophrenic monozygotic twins. European Archives of Psychiatry & Clinical Neuroscience 1993; 243:7-10.

Rubinstein G. Schizophrenia, rheumatoid arthritis and natural resistance genes. Schizophrenia Research 1997; 25:177-81.

Schwarz MJ, Chiang S, Muller N et al. T-helper-1 and T-helper-2 responses in psychiatric disorders. Brain, Behavior, & Immunity 2001; 15:340-70.

Sharpley M, Hutchinson G, McKenzie K et al. Understanding the excess of psychosis among the African-Caribbean population in England. Review of current hypotheses. British Journal of Psychiatry - Supplementum 2001; 40:s60-8.

Strassburg CP, Obermayer-Straub P, Manns MP. Autoimmunity in hepatitis C and D virus infection. Journal of Viral Hepatitis 1996; 3:49-59.

Susser ES, Lin SP. Schizophrenia after prenatal exposure to the Dutch hunger winter of 1944- 1945. Archives of General Psychiatry 1992; 49:983-8.

Swedo SE, Leonard HL, Mittleman BB et al. Identification of children with pediatric autoimmune neuropsychiatric disorders associated with streptococcal infections by a marker associated with rheumatic fever [see comments]. American Journal of Psychiatry 1997; 154:110-2.

Torrey EF. Are we overestimating the genetic contribution to schizophrenia? Schizophrenia Bulletin 1992; 18:159-70.

Torrey EF, Miller J, Rawlings R et al. Seasonality of births in schizophrenia and bipolar disorder: a review of the literature. Schizophrenia Research 1997; 28:1-38.

Torrey EF, Yolken RH. At issue: is household crowding a risk factor for schizophrenia and bipolar disorder? Schizophrenia Bulletin 1998; 24:321-4.

Torrey EF, Yolken RH. Could schizophrenia be a viral zoonosis transmitted from house cats? Schizophrenia Bulletin 1995; 21:167-71.

Torrey EF, Yolken RH. Familial and genetic mechanisms in schizophrenia. Brain Research Reviews 1999; 31:113-7.

Torrey EF, Yolken RH. The schizophrenia-rheumatoid arthritis connection: infectious, immune, or both? Brain, Behavior, & Immunity 2001; 15:401-10.

Tsuang M. Schizophrenia: genes and environment. Biological Psychiatry 2000; 47:210-20.

Yolken RH, Bachmann S, Rouslanova I et al. Antibodies to Toxoplasma gondii in individuals with first-episode schizophrenia. Clinical Infectious Diseases 2001; 32:842-4.

Chapter 5

THE ROLE OF VIRAL INFECTIONS DURING NEURODEVELOPMENT

In recent years substantial evidence has accumulated to support the contention that schizophrenia has its origins in abnormal brain development. Viruses are prominent candidates as the presumed environmental component of this process, acting in concert with susceptibility genes to bring about an aberrant developmental trajectory that is ultimately manifested as psychotic symptoms. This chapter will elucidate the points at which neurodevelopmental and viral mechanisms intersect, and provide a critical overview of this sub-theory of the viral hypothesis, which postulates that at least some cases of schizophrenia are caused by a viral infection during gestation, or in the early postnatal years.

Viruses are established CNS teratogens that can have devastating effects on the developing brain as exemplified by congenital rubella or cytomegalovirus. What's less well understood is the extent to which viruses can invade the developing brain and cause subtle alterations in neural circuitry. It has also been difficult to define in humans the various indirect ways that maternal immune molecules can disrupt fetal neurodevelopment in the absence of fetal CNS infection. To further complicate matters, many viral infections do not come to clinical attention even when they occur during pregnancy, and only in rare cases involving severe symptoms or clear diagnostic signs is the responsible virus isolated or identified.

The fetus is vulnerable to ascending infections of the genital tract (particularly if fetal membranes ruptured), as well as viruses that cross the placenta (Kaplan 1993; Nahmias et al. 1997). Once the fetus becomes infected, its immature blood brain barrier may not offer adequate defense against viral invasion or pernicious immune molecules. In the case of schizophrenia, the established hallmarks of a viral infection, such as inclusion bodies or viral antigens may be undetectable at autopsy because the infection could insidiously disrupt development and then be cleared from the brain prior to the manifestation of symptoms decades later. Antibody serology might also fail to detect past evidence of the infection because immunoglobulin production develops slowly during gestation and remains impaired even at birth. These are important obstacles to elucidating the contribution of viral infections to adult schizophrenia, and should be kept in mind when evaluating the available data concerning the role of virus in the neurodevelopmental origins of adult psychopathology.

The first two sections of this chapter provide a brief overview of normal brain development and fetal-maternal immunology. The neurodevelopmental hypotheses for schizophrenia, is then explored including evaluation of increasingly convincing evidence that suggests a role for viruses in the concatenation of adverse events eventually leading to psychosis.

5.1 BRAIN DEVELOPMENT

Brain development is a magnificently programmed event that unfolds under the constant influence of the intrauterine and postnatal environment. There are some fundamental rules how this program unfolds (Figure 1). These rules are mostly common sense: e.g. because time does not flow backwards a later event cannot alter the outcome of an earlier event. This is not to say that the disruption of an early event cannot, in some manner, be compensated for by a later occurring event. Since the histology of human development is discerned as a series of snapshots, the time of occurrence of an observable neuromorphological anomaly is deduced from the known sequence of normative developmental, and in some cases, through extrapolation from the known timing of an environmental insult. This process of deduction is particularly difficult when the neurohistological examination used to date the insult must rely on adult autopsy tissue, as is usually the case in schizophrenia.

The nervous system is initially formed from predominantly ectodemal cells in a process called neurulation, which occurs in the first three to five weeks after conception. This process results in the formation of the neural tube from which the central nervous system is derived. Major defects at this early stage are not thought to be related to schizophrenia since they usually result in spontaneous abortion. As shown in figure 2, much of cerebral development takes place between the 3rd and 7th months in utero. Many neurodevelopmental theories of schizophrenia focus on this mid-gestation period, though the responsible insult could certainly occur at other stages, or even postnatally.

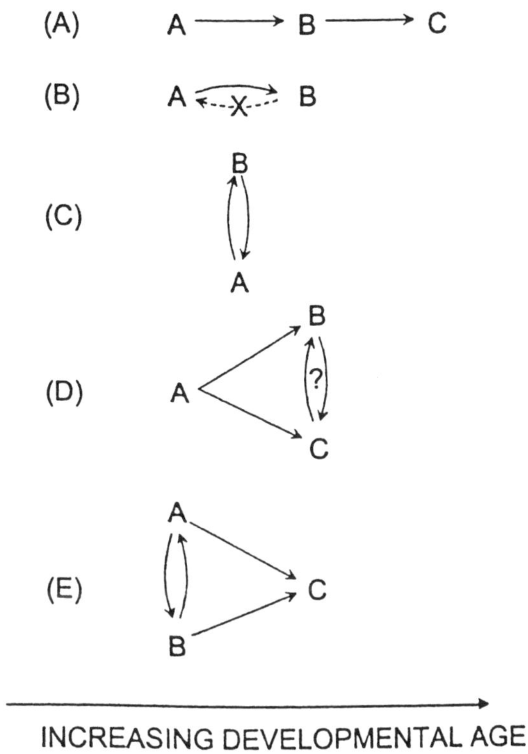

Figure 1 . Examples of how events during development influence each other in different ways. (A) Events which follow each other in a sequence can be causally related. (B) A late-occurring event cannot influence an event occurring earlier. (C) Two events occurring at the same time can influence each other. (D) One early-occurring event can influence two (or more) events occurring later, which may or may not influence each other. (E) One event may be dependent on the interaction of two (or more) events occurring earlier. Reproduced with permission from: Nowakowski, R. & Hayes, N. CNS Development: An overview. Development and Psychopathology 1999; 11: 398.

Brain development is best conceptualized as a series of overlapping events rather than discreet stages (Nowakowski et al. 1999). These events are diagrammed in figure 3. The various brain regions develop according to their own timetables, and consequently there is extensive overlap in these cellular and molecular events that eventually give rise to a mature brain. For example, the major nuclei of the mesencephalic tegmentum are mostly in place by the first six to seven weeks gestation, whereas formation of the hippocampus has just began at this time.

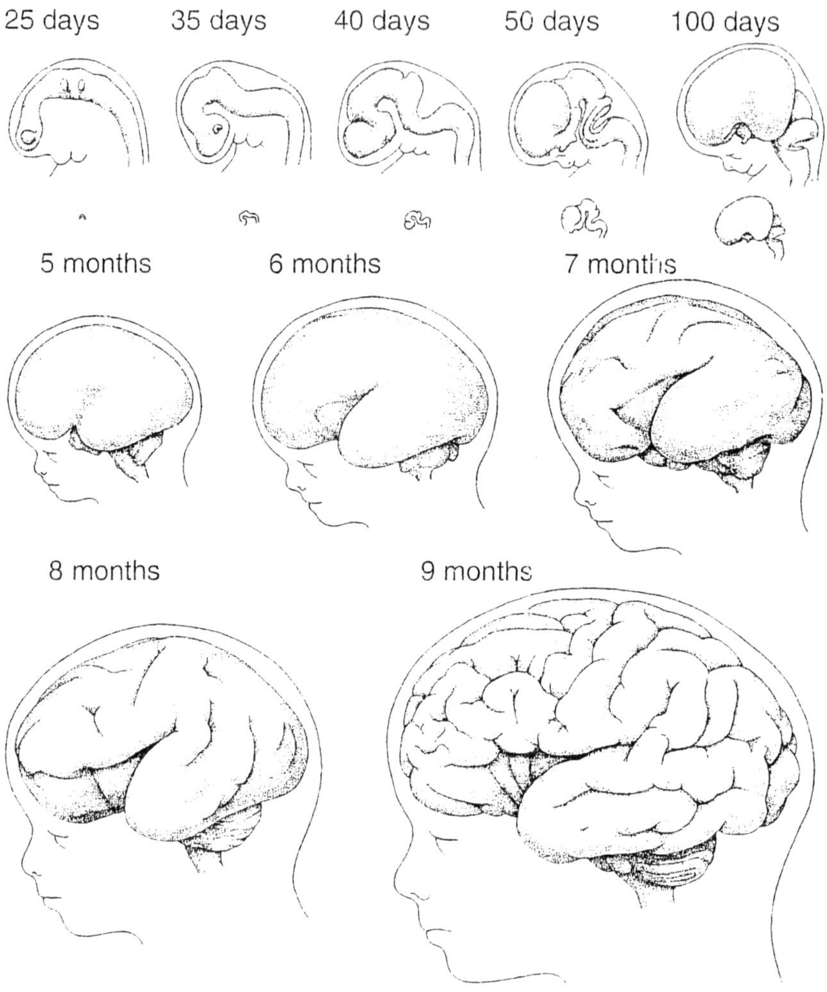

Figure 2. Developing human brain is viewed from the side in this sequence of drawings, which show a succession of embryonic and fetal stages. (Reproduced with permission from Cowen, W.M., Scientific American, Volume 241, 1977)

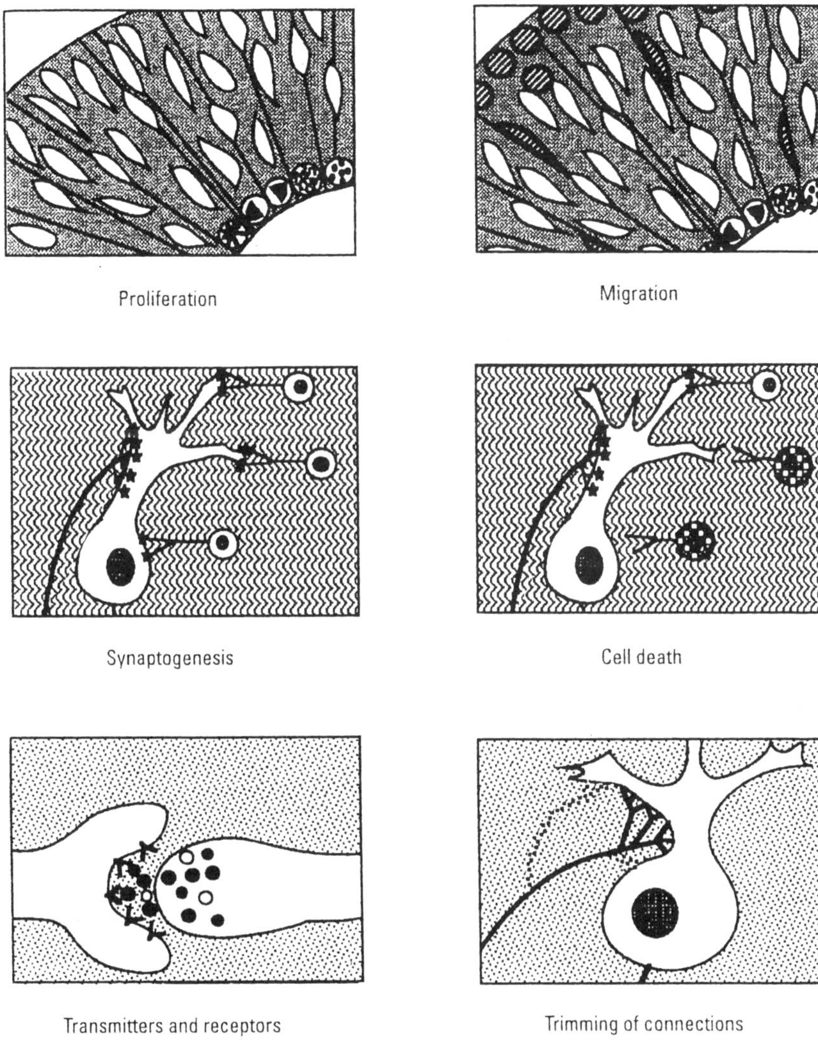

Figure 3. Developmental processes. Reproduced with permission from: Rodier, P. Developing Brain as a Target of Toxicity. Environmental Health Perspectives 1995; 103: 74.

Cell proliferation occurs as series of waves with different timing in different brain regions. New neurons are added even before the neural tube closes, and continue to be added in some brain regions such as the hippocampus well into adult life. The formation of new neurons is called "neurogenesis." Insults that impair neurogenesis will decrease the number of neurons being formed at the time of the insult.

Neurons are not usually produced at the site of their final destination and they must travel or migrate in order to establish proper connections and cytoarchitecture (Arnold 1999; Nowakowski & Hayes 1999). In some brain regions neurons are pushed into their position rather passively by the proliferation of their younger neighbors. In other brain regions neurons travel by a more active process, often guided by glial fibers. An insult that occurs during the migration and settling of neurons will result in an ectopia--the presence of misplaced cells.

As newborn neurons mature and differentiate they form connections with other neurons in a event referred to as synaptogensis. It is now clear that synaptogensis is not limited to development, but occurs constantly as the mature brain is remodeled through experience. However, the formation of synapses and elaboration of neurotransmitters has a particularly profound effect on the ultimate neural circuitry of the developing brain, in part because there is an over-production of neurons during development, and these become trimmed-back based on their connections as some undergo apoptosis, and others become permanent members of the neural circuitry. In this way neurons that fail to produce the correct neurotransmitters or make the appropriate connections may demonstrate abnormalities in their axons or dendritic arbor, or these cells may die. The brain also produces an over-abundance of synaptic connections during gestation. These are selectively eliminated postnatally (Feinberg 1982). The encasement of axonal fibers by myelin also continues long after birth, and even into adulthood.

By the time of birth the brain has grown to 350 grams, but most brain growth occurs after birth as the of the brain expands to its adult size of 1300 grams. Thus, based on weight, the growth of the brain can be viewed as a continuous process beginning soon after conception and transitioning through parturition in a seamless journey to its mature size many years later. However, most of the structural organization of the brain is laid out during gestation, and importantly, this organization and construction is occurring in the unique milieu of the womb, and hence neuromorphogenesis depends on maternal physiology and immunology.

5.2 IMMUNOLOGY OF THE MATERNAL-FETAL UNIT

The immunological relationship between a fetus and its mother is unique. Because the fetus and mother do not share the same versions of MHC molecules, the fetus might be expected to be recognized by the mother as a foreign tissue, an allograft (Nahmias & Kourtis 1997). The mechanisms that come into play and prevent the mother from rejecting the fetus are not completely understood, but the placenta, which forms the interface between fetal and maternal circulation, plays an important role. Another consequence of the mismatch between the fetal and maternal MHC is that the mothers' acquired cellular immunity cannot operate within the fetal compartment, and therefore the immunity the mother has built up over time can only be transferred to the fetus by soluble molecules such as antibodies.

The fetus obviously does not begin with an intact immune system at conception. The cells and tissues of the immune system develop during prenatal and postnatal ontogeny; this process cannot help but be influenced by environmental factors and maternal health (Coe et al. 2000).

The fetal immune system has to learn to develop immune tolerance to the multitude of proteins and other molecules that become newly existent during the process of assembling a living human. The task of distinguishing these nascent self-molecules from foreign microbial antigens requires that the fetus perform a delicate balancing act during immune system ontogeny. Consequently, the fetal response to infection is most often quantitatively and qualitatively different than the response would be in an adult.

Some immune cells of the developing fetus begin to be produced surprisingly early; T-cells can be detected by six weeks gestation and by two months lymphocytes have entered the thymus. The expression of MHC class I and II is evident in many fetal tissues by three months of gestation, and the presence of lymph nodes, thymic tissue, and B cells at this stage suggest that a primordial immune response is intact by the beginning of the second trimester. Even so, almost all aspects of the fetal immune system are deficient compared to adults at this stage. While the fetus can make some limited B cell responses to infection starting late in the second trimester, it is largely dependent on passively-transferred antibody from its mother for protection against viral infections (Zinkernagel 2001). Maternal antibody enters the fetal circulation at the beginning of the first trimester and peaks at approximately the time of birth. The fetal contribution to circulating IgG is virtually nil until the first month or so after birth. Indeed, the B cell response does not reach adult equivalents until several years into childhood. T cell function is also impaired in the fetus and neonate compared to adults. The fetus and neonate have an greater proportion of naive B- and T- cells than adults because of the unlikelihood that the fetus would encounter antigens that it could recognize as a foreign during the a tolerant state of immune system ontogeny.

The immature immune system of the fetus or neonate appears to be particularly refractory to certain antigens, particularly polysaccharides. This may serve the purpose of preventing the production of autoantibodies against certain brain constituents. For example meningococcus B has a capsule composed of polysialic acid that is identical to the polysialic acid on the brain neural cell adhesion adhesion molecule, PSA-NCAM (Nedelec et al. 1990). If a fetus made an antibody response against the polysaccharide of the meningococcus bacterium while the blood brain barrier was immature, this would open an opportunity for auto-reactive antibodies to disrupt development of structures expressing PSA-NCAM. Thus the "blind spot" for recognizing and responding to certain polysaccharide antigens actually serves a neuroprotective functions for the immature brain.

Besides antibodies, cytokines are another class of molecules that permit immunological interchange between the mother and a fetus. There is burgeoning interest among obstetrical researchers in tracking changes in cytokines as indices of infectious events during pregnancy (Romero et al. 1998). Numerous reports have suggested that proinflammatory cytokines, IL-6 in particular, are generated at high levels during intrauterine infection and may be critical triggers of premature labor. There is also an accumulating body of evidence that cytokines, particularly those produced during infections, play a role in the pathophysiology of perinatal brain damage (Dammann et al. 2000). The potential role of maternal cytokines in the neurodevelopmental irregularities of schizophrenia is an important area that has just recently gained attention (Urakubo et al. 2001).

The placenta allows for maternal support of fetal circulation and survival. There is direct contact between maternal blood and fetal tissue as maternal blood spills into the intervillus space into which fetal tissue villi project. Thus the placenta is composed of maternal and fetal cells, and a genetic defect on either the maternal or fetal side could influence the ingress of immune molecules into the fetal compartment. Considering the large number of immunoreactive substances, including cytokines, that are localized in the placenta, this tissue is clearly a site of immunological cross-talk capable of influencing the homeostasis of both mother and fetus.

5.3 THE VIRAL-DEVELOPMENTAL HYPOTHESIS

The neurodevelopmental hypothesis of schizophrenia, which implies latency between the initial insult and the manifestation of symptoms, remains the prevailing paradigm for understanding schizophrenia pathogenesis (Weinberger, 1995). As discussed above, the precise timing of an insult that disrupts fetal development determines the character of the pathology and the brain region affected. In theory the insult could occur at any time in development, even soon after the neural tube formed if the virus was sufficiently non-cytopathic. Table 1 shows the timing of some overlapping events in neurodevelopment and proposes the type of pathological consequences that would be incurred due to an insult coinciding with these developmental stages. Discovering the nature of the "initial insult" in schizophrenia has become an urgent goal of biological psychiatry, and as alluded to in the previous chapter, there may be manifold types of fetal disruptions underlying the disease, but viruses warrant prominent consideration.

A neurodevelopmental origin of schizophrenia has been formulated by numerous investigators and is supported by several converging lines of evidence derived from studies in neuropsychology, developmental biology, and pathology (Bloom 1993; Murray et al. 1992; Walker et al. 1999; Weinberger 1987). Neuropsychological studies have shown that while schizophrenia is not typically manifested clinically until late adolescence, behavioral and motor signs of risk are observed much earlier indicating the existence of an insidious brain disorder that precedes psychotic symptoms. Minor physical malformations such as craniofacial asymmetries or dermatoglyphic irregularities are also increased in patients with schizophrenia suggesting an association with aberrant morphogenesis. These physical abnormalities occurring outside the brain are considered nonspecific indicators of schizophrenia and are not thought to be involved in disease causation; however, from the perspective of a viral etiology this apparently reasonable assumption cannot be taken for granted because subtle defects in peripheral development could include immune tissues (e.g., tonsillar lymphatics, thymus) and thus alter viral susceptibility or immune responses.

Cytoarchitectural studies also support the neurodevelopmental hypothesis, and suggest that disruption of neuronal circuits during development is an important antecedent to schizophrenia (Arnold 1999; Bloom 1993; Murray et al. 1992; Weinberger 1987). Autopsy studies of schizophrenic brains have found evidence of disturbed neuronal formation and migration, abnormal expression of developmentally-regulated molecules such as NCAM, and a lack of gliosis in brain regions exhibiting cell loss. Of note, abnormal levels of an NCAM isoform have been found in the CSF of schizophrenia patients. This abnormality did not appear to be not linked to antipsychotic medication (van Kammen et al. 1998), and because monozygotic twin pairs discordant for schizophrenia where likewise discordant for these CSF NCAM elevations, an environmental insult during neurodevelopment could have triggered persistent alterations in this developmentally-regulated molecule (Poltorak et al. 1997).

Current neurodevelopmental theories of schizophrenia posit that both genetic and environmental factors contribute to aberrant brain maturation (Tsuang 2000). However, as discussed further in chapter 6, there are many possible models incorporating a viral infection at some stage that could explain the available information on schizophrenia pathogenesis. Thus it is conceivable that schizophrenia could be a disorder which, like Alzheimer's disease, has purely genetic subtypes as well as "sporadic" forms which arise due to environmental agents acting on genetically susceptible individuals.

Neurodevelopmental Process	Approximate Timing[a]	Possible Related Neuropathological Abnormalities
Neurogenesis	E4 weeks–E32 weeks	Atrophy Neuron number
Neuronal migration	E4 weeks–P4 months	Cytoarchitecture Neuron orientation Number and distribution of white-matter neurons Guidance molecules (e.g., N-CAM, reelin)
Programmed cell death	E6 weeks–P6 months	Neuron number Cytoarchitecture Number and distribution of white-matter neurons Trophic factors
Synaptogenesis	E20 weeks–childhood	Synaptic density Synapse-related protein and mRNA expression Trophic factors
Dendritic and axonal arborization	E22 weeks–adulthood	Neuron density Neuron size Cytoskeletal protein expression (e.g., MAP2) Dendritic length and branching Dendritic spine number and clustering patterns Growth cone protein expression Trophic factors
Myelination	E23 weeks–adulthood	No studies to date
Synaptic elimination	P4 months–adulthood	Synaptic density Synapse-related protein and mRNA expression

[a]E, embryonic age; P, postnatal age.

Table 1 Neurodevelopmental events and possible related neuropathological abnormalities that have been described in schizophrenia. Reproduced with permission from: Arnold, S. Neurodevelopmental abnormalities in schizophrenia: Insights from neuropathology. Development and Psychopathology 1999; 11:4 43.

The ability of environmental factors to affect brain structures implicated in schizophrenia is underscored by a recent study of monozygotic twins who were discordant for schizophrenia; hippocampal volumes were consistently reduced only in the afflicted members of the pairs (Suddath et al. 1990). Studies pertaining to the intrauterine environment of twins are informative in this regard. Monozygotic twins can either share the same placenta, in which case they are termed "monochorionic, or each twin can have its own placenta (and thus its own blood circulation) in which case the pairs are termed "dichorionic" (Figure 4). Concordance rates for schizophrenia are higher for monochorionic versus dichorionic twins (Davis et al. 1995). Since in either case the twin pairs begin with essentially the same genome, the higher incidence of schizophrenia in monochorionic twins is a function of their shared placenta, and is consistent with the idea that a virus or other pathogen causes schizophrenia by infecting the placenta and invading the enveloped fetus(es). The observed concordance rates for twins could also be explained within the framework of a virus-triggered autoimmune reaction because monozygotic twins have different T- and B- cell repertoires, and thus a virus could infect both twins equally but the immune response could be the essential determinant of disease expression (Squires 1997).

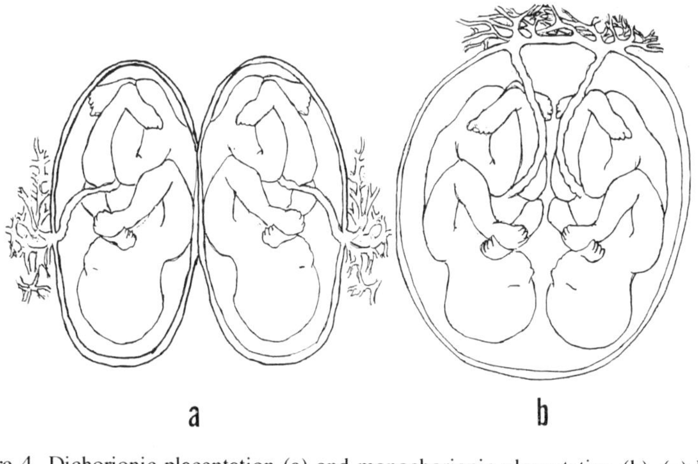

Figure 4 Dichorionic placentation (a) and monochorionic placentation (b). (a) Illustrates the separate placentas and separate fetal circulations of dichorionic twins, who can be either dizygotic or monozygotic. Figure 4 b illustrates the single placenta and shared vascular communications of monochorionic placentation, which occurs only in monozygotic twins. Reproduced with permission from: Davis, J., Phelps, J., & Bracha, H. Prenatal Development of Monozygotic Twins and Concordance for Schizophrenia. Schizophrenia Bulletin 1995; 21 (3): 358.

5.4 EPIDEMIOLOGICAL STUDIES OF PERINATAL INFECTIONS IN ADULT SCHIZOPHRENIA: THE RETROSPECTIVE DILEMMA

The age of onset for positive symptoms of schizophrenia peaks between 17 and 27 years for males, and 17 to 37 years for females. According to the neurodevelopment hypothesis, the insult leading to these symptoms would have taken place about 2 to 3 decades earlier. Because the neurodevelopmental origin for the disease has only recently gained favor among schizophrenia experts, epidemiological studies of developmental risk factors have had to rely largely on retrospective studies. Worse yet, the idea of viruses as potential causal factors in schizophrenia was largely dismissed until the last 15 years. Nevertheless, epidemiologists have devised ingenious ways to circumvent the historical bias against the viral hypothesis, and have found intriguing and compelling associations between schizophrenia and several gestational infections.

5.4.1 Influenza

In utero exposure to influenza has been the most extensively studied infectious antecedent of schizophrenia (Watson et al. 1999). This is in part for practical reasons: influenza is relatively common in developed countries, and these countries have the means and the will to track infectious incidence and record epidemics. This makes it possible to study a large number of people with respect to influenza exposure and subsequent schizophrenia.

Although Menninger had considered the possibility that influenza might cause schizophrenia by acting near the time of the psychotic episode, the link with neurodevelopment was suggested by the ground breaking study of Mednick and colleagues that found an implied association between maternal influenza and the later development of schizophrenia in the offspring (Mednick et al. 1988). Individuals exposed to the 1957 Helsinki influenza epidemic in their second trimester of gestation were found to have an increased likelihood of being diagnosed with schizophrenia upon being assessed more than 25 years later. Because previous studies had shown that exposure to influenza during the first trimester of gestation could result in a relatively overt CNS anomalies, Mednick et al. hypothesized that "... later exposure, in the second or third trimester of fetal life, would be associated with more subtle CNS damage that would not be observable in the form of gross birth anomalies but would increase the risk of adult schizophrenia" (Mednick et al. 1988).

The findings of Mednick et al. have been replicated, or partially replicated (some studies found the effect limited to female fetuses) for mid-gestation exposure to influenza epidemics in several countries including the United Kingdom, Scotland, Denmark, Japan, and Australia (Brown et al. 2002). However, other attempts to replicate the findings in the United States, Netherlands, Croatia, Ireland, and the United Kingdom have not been successful.

The aforementioned studies examined epidemics in a single year (e.g. 1957 in Helsinki). Another way to examine influenza exposure is to relate schizophrenia births to the annually varying prevalence of influenza epidemics. There have been 10 such studies, the majority of which support Mednick's hypothesis.

One of the criticisms of the original influenza study, and most of the replication attempts, is that it did not actually trace back individual patient histories to determine if the mother actually had influenza or any other disease during her pregnancy. Mednick and his colleagues addressed this question in a subsequent study by examining obstetric records of the schizophrenia proband's gestational history; the association with maternal influenza during the second trimester remained valid (Mednick et al. 1994).

However, there have been two cohort studies that have not found a connection between maternal influenza and schizophrenia. In one study there was no relationship between maternal influenza infection, as judged by a midwife records, and the incidence of schizophrenia in the offspring (Crow et al. 1992). Another study followed up mothers who claimed to have influenza during pregnancy (Cannon et al. 1996). Specifically, around the time of delivery (in 1957-58), the mothers had been posed the question: " have you ever had influenza during this pregnancy." When their children from this pregnancy were followed up three decades later there was no association between a mother's affirmative answer to this question, and the later development of schizophrenia in the child. Despite the inherent appeal of this study's design, there are some problems. Just because a woman claimed that she had influenza does not mean she actually was infected with the virus; and conversely, some women could have been infected and had only minor symptoms because of a vigorous primary or secondary immune response. Accordingly, if the antibody response, rather than the virus itself, was responsible for perturbing brain development then the children of the latter group would be at the highest risk for schizophrenia. Indeed, maternal recall is an issue: other studies have shown that about a quarter of those women who believed they suffered an influenza infection lacked detectable antibodies. On the other hand, 60% of women who denied symptoms of influenza during the 1957 epidemics had serological indices of a recent influenza infection (Arvin et al. 1990).

One possible explanation for the discrepancies in the influenza-schizophrenia literature is that influenza may be a surrogate marker for the actual microbial culprit, which could cycle in synchrony with influenza, or alternatively could be a non-cyclical endemic agent that was nonetheless able to take opportunistic advantage of patients with influenza. In this regard, prenatal or postnatal infection with a number of other viruses or bacteria has been implicated epidemiologically as a risk factor for adult schizophrenia. These included Varicella-zoster, measles, and polio (O'Callaghan et al. 1994; Rantakallio et al. 1997; Suvisaari et al. 1999; Torrey 1988).

A recent study used prospectively-collected data on maternal respiratory infections to examine the impact of second trimester exposure on the later development of a schizophrenia spectrum disorder (Brown et al. 2000b). Most of the infections in this study were documented by a physician, and the cohort was followed continuously from 1981 to 1997. For statistical reasons, the various respiratory infections were diagnostically-collapsed into a single category. Second trimester exposure was found to be significantly associated with an increased risk for a schizophrenia spectrum disorder. This association remained significant even after adjusting for maternal smoking.

5.4.2 Rubella and Herpes Simplex

In their study of rubella, Brown and Susser utilized a birth cohort in which the members were prospectively determined to have been exposed to rubella virus in utero (Brown et al. 2000a; Brown & Susser 2002). For most of the cases congenital infection was documented by serological analysis.

Congenital rubella has long been known to cause birth defects, including anomalies of the CNS. However, the characteristics that make rubella most germane to schizophrenia are its ability to persist in nervous tissue, and the tremendous variability in fetal outcome following a congenital infection (Volpe 1995). Intra-uterine exposure to rubella can result in miscarriage or severe multi-organ defects. However, approximately 70% of infected newborns have no evident anomalies. There is considerable data to show that these silent infections may become manifested in childhood as neurodevelopmental delays, sensory deficits, or even autism. Studies of twins have shown that monozygotic twins can be concordant or discordant for rubella infection, though there appears to be a significant genetic component to susceptibly.

Brown and Susser studied these rubella-exposed individuals who were spared serious birth defects --including mental retardation-- despite the fact that they were in utero at the time of maternal rubella (Brown et al. 2000a; Brown & Susser 2002). These children were assessed for mental illness at 21 to 23 years old, and again at 33 to 35 years old. The exposed children had a dramatically increased risk of developing a schizophrenia spectrum disorder. In the individuals assessed at 21 to 23 years old, the risk ratio for this diagnosis was between 5.2 and 16.3, depending on the comparison group. By the time that cohort members were 33 years old, 20% of them had been diagnosed with one of the schizophrenia spectrum disorders.

This study is compelling and represents a major advance in the epidemiological investigation of the viral hypothesis. Although rubella is rare in current times because of effective vaccination, the findings of this study indicate that a neurotropic virus can cause subtle perturbations in development that are eventually manifested as a psychotic disoder, including schizophrenia. The mechanism by which rubella may exert a latent disruption of brain function is not known, though many of the virus's effects on the CNS and other organs are thought to involve inhibition of cell division. Interestingly, we have used an animal models to demonstrate that a virus acting during development can cause a latent loss of neurons in the hippocampus (Sharma et al. 2001). This effect appears to be mediated through the destruction or impairment of pluripotent neuronal progenitors cells that give rise to these neurons.

Herpes simplex virus is another virus that has been investigated using a prospective cohort approach (Buka et al. 2001). However, in this case infection with the virus was not directly documented. The study used archival maternal plasma to test the hypothesis that maternal infections were associated with giving birth to child who later developed a psychotic disorder. The mother's plasma had been banked as part of the Collaborative Perinatal Project, which was assembled between 1959 and 1966 in order to identify perinatal factors influencing child development. Initial assays revealed an elevation of the total serum IgG and IgM (usually interpreted as a nonspecific indication of recent infection) in the mothers of children who eventually developed psychosis compared to the mothers of control children-who were matched for sex, ethnicity, and the season of birth. Additional serological examination of antibody titers against several infectious diseases suggested that mothers of the cohort members with psychosis had higher titers against HSV-2 near the time of delivery than the mothers of controls.

There is no way of telling from this study whether at some point during the pregnancy the mother had a primary HSV-2 infection, which is most likely to be transmitted to the neonate, or if the higher titers to HSV-2 indicated a reactivation of the virus, or if the higher titers reflect a generally more vigorous immune response against a pre-existing HSV-2 infection. The study found a graded association between maternal anti-HSV-2 antibodies and adult psychosis, which might argue for the latter possibility in that it suggests that the amount of antibody, and not seropositivity per se, brings about an adverse fetal outcome. Moreover, because the mothers most likely would have contracted the infection through sexual contact, the "infected" mothers with elevated anti-HSV-2 IgG might be expected to have evidence of infection with other sexually transmitted diseases. Additional serological testing indicated this was not the case (Buka et al. 2001). Thus, its possible that the mothers were generating antibodies against an as yet undiscovered herpes virus, and the assay was detecting cross-reacting epitopes. Still, an elevated IgG against HSV-2 could indicate that the virus was being reactivated near the time of birth. This raises the question of whether congenital infection with HSV could remain silent in the neonate, and travel to the brain to establish an occult infection capable of redirecting neuromorphogenesis toward an ultimate schizophrenia phenotype.

Neonatologists recognize congenital herpes as a serious disease that is almost always symptomatic and can have fatal consequences (Nahmias et al. 1967; Whitley 1997). However, the identification of congenital HSV-2 infection is not uniform throughout the world, and there are indications that the diagnosis could be missed, even in developed countries. When the virus invades the brain it's generally assumed to cause a severe encephalitis in which permanent neurological impairment is the rule. But even in the infants for whom the virus appears to remain localized to the integument, neurological dysfunction eventually develops in about 30% of the cases. Thus it is possible that HSV-2 skin lesions might be misdiagnosed (especially in the early 1960's when the blood in the above study was drawn) and some children in the aforementioned study could have developed neuropsychiatric sequela due to an earlier subclinical invasion of the brain by the virus. This not withstanding, the incidence of congenital herpes is likely too low (1 in 5000 births) to account for a large number of schizophrenia cases.

In spite of these caveats, there is no doubt that complications in neurodevelopment produced by congenital infections can range broadly in severity. For example, 90% of neonates infected prenatally with CMV are asymptomatic. Of these, about 15% will develop latent neuropsychiatric complications when followed longitudinal through childhood. The consequences for the remaining 85% who had silent infections has not been well studied.

5.4.3 Some Solutions to the Retrospective Dilemma

Besides establishing better cohorts, and continuing to follow those that have been established, there are additional ways to approach the long latency between the presumed infection and eventual diagnosis of schizophrenia. Identifying schizotypal patients and performing serial testing of antibody titers along with MRI scans is one such approach we have recently considered. This might help determine if there is reactivation of a virus associated with deterioration in psychological functioning or specific regional changes in brain anatomy. Still, because brain development is a dynamic process involving interacting and interdependent neuronal elements, extrapolation of human data is an ineffective means to determine the brain region(s) in which the original ontogenetic abnormality occurred.

Animal models using viral infections precisely timed to specific periods of neurodevelopment have the potential to reveal specific interconnected sets of vulnerable neurons, and perhaps provide clues to disentangling the pathophysiological underpinnings of schizophrenia subtypes. Indeed, one of the strengths of the viral hypothesis for schizophrenia is the well know ability of viruses to target specific types of neurons in the CNS. Conversely, interdependent host factors such as the developmental stage at infection, expression of viral entry receptors, and immune responses also determine the cell types infected and the ultimate pathology engendered by the virus.

An example of such pathology is aberrant neuronal migration, which has been demonstrated in experimental hosts after prenatal or neonatal infection with influenza virus, murine cytomegalovirus, mumps virus and parvovirus. The mechanisms by which some viruses interfere with the cellular machinery governing neuronal migration has not been well characterized and is highly amenable to study in animal models.

REFERENCES

Arnold SE. Neurodevelopmental abnormalities in schizophrenia: insights from neuropathology. Development & Psychopathology 1999; 11:439-56.

Arvin AM, Yeager AS.Other infections of the fetus and newborn. In: JS Remington; JO Klein, editors Infectious diseases of the fetus and newborn. Philadelphia: W.B. Saunders; 1990.

Bloom FE. Advancing a neurodevelopmental origin for schizophrenia. Archives of General Psychiatry 1993; 50:224-7.

Brown AS, Cohen P, Greenwald S et al. Nonaffective psychosis after prenatal exposure to rubella. American Journal of Psychiatry 2000a; 157:438-43.

Brown AS, Schaefer CA, Wyatt RJ et al. Maternal exposure to respiratory infections and adult schizophrenia spectrum disorders: A prospective birth cohort study. Schizophrenia Bulletin 2000b; 26:287-95.

Brown AS, Susser ES. In utero infection and adult schizophrenia. Mental Retardation & Developmental Disabilities Research Reviews 2002; 8:51-7.

Buka SL, Tsuang MT, Torrey EF et al. Maternal infections and subsequent psychosis among offspring. Archives of General Psychiatry 2001; 58:1032-7.

Cannon M, Cotter D, Coffey VP et al. Prenatal exposure to the 1957 influenza epidemic and adult schizophrenia: a follow-up study. [see comments]. British Journal of Psychiatry 1996; 168:368-71.

Coe CL, Lubach GR. Prenatal influences on neuroimmune set points in infancy. Annals of the New York Academy of Sciences 2000; 917:468-77.

Crow TJ, Done DJ. Prenatal exposure to influenza does not cause schizophrenia. British Journal of Psychiatry 1992; 161:390-3.

Dammann O, Leviton A. Role of the fetus in perinatal infection and neonatal brain damage. Current Opinion in Pediatrics 2000; 12:99-104.

Davis JO, Phelps JA, Bracha HS. Prenatal development of monozygotic twins and concordance for schizophrenia [published erratum appears in Schizophr Bull 1995;21(4):539]. Schizophrenia Bulletin 1995; 21:357-66.

Feinberg I. Schizophrenia: caused by a fault in programmed synaptic elimination during adolescence? Journal of Psychiatric Research 1982; 17:319-34.

Kaplan C. The placenta and viral infections. Seminars in Diagnostic Pathology 1993; 10:232-50.

Mednick SA, Huttunen MO, Machon RA. Prenatal influenza infections and adult schizophrenia. Schizophrenia Bulletin 1994; 20:263-7.

Mednick SA, Machon RA, Huttunen MO et al. Adult schizophrenia following prenatal exposure to an influenza epidemic. Archives of General Psychiatry 1988; 45:189-92.

Murray RM, Jones P, O'Callaghan E et al. Genes, viruses and neurodevelopmental schizophrenia. Journal of Psychiatric Research 1992; 26:225-35.

Nahmias AJ, Josey WE, Naib ZM. Neonatal herpes simplex infection. Role of genital infection in mother as the source of virus in the newborn. JAMA 1967; 199:164-8.

Nahmias AJ, Kourtis AP. The great balancing acts. The pregnant woman, placenta, fetus, and infectious agents. Clinics in Perinatology 1997; 24:497-521.

Nedelec J, Boucraut J, Garnier JM et al. Evidence for autoimmune antibodies directed against embryonic neural cell adhesion molecules (N-CAM) in patients with group B meningitis. Journal of Neuroimmunology 1990; 29:49-56.

Nowakowski RS, Hayes NL. CNS development: an overview. Development & Psychopathology 1999; 11:395-417.

O'Callaghan E, Sham PC, Takei N et al. The relationship of schizophrenic births to 16 infectious diseases. British Journal of Psychiatry 1994; 165:353-6.

Poltorak M, Wright R, Hemperly JJ et al. Monozygotic twins discordant for schizophrenia are discordant for N-CAM and L1 in CSF. Brain Research 1997; 751:152-4.

Rantakallio P, Jones P, Moring J et al. Association between central nervous system infections during childhood and adult onset schizophrenia and other psychoses: a 28-year follow-up. International Journal of Epidemiology 1997; 26:837-43.

Romero R, Gomez R, Ghezzi F et al. A fetal systemic inflammatory response is followed by the spontaneous onset of preterm parturition. American Journal of Obstetrics & Gynecology 1998; 179:186-93.

Sharma AK, Valadi N, Miller AH et al. Adult neurogenesis in the hippocampus is impaired following neonatal viral infection. Neurobiology of Disease 2001; Under Review.

Squires RF. How a poliovirus might cause schizophrenia: a commentary on Eagles' hypothesis. Neurochemical Research 1997; 22:647-56.

Suddath RL, Christison GW, Torrey FE et al. Anatomical abnormalities in the brains of monozygotic twins discordant for schizophrenia. The New England Journal of Medicine 1990; 322:789-94.

Suvisaari J, Haukka J, Tanskanen A et al. Association between prenatal exposure to poliovirus infection and adult schizophrenia. American Journal of Psychiatry 1999; 156:1100-2.

Torrey EF. Stalking the schizovirus. Schizophrenia Bulletin 1988; 14:223-9.

Tsuang M. Schizophrenia: genes and environment. Biological Psychiatry 2000; 47:210-20.

Urakubo A, Jarskog LF, Lieberman JA et al. Prenatal exposure to maternal infection alters cytokine expression in the placenta, amniotic fluid, and fetal brain. Schizophrenia Research 2001; 47:27-36.

van Kammen DP, Poltorak M, Kelley ME et al. Further studies of elevated cerebrospinal fluid neuronal cell adhesion molecule in schizophrenia. Biological Psychiatry 1998; 43:680-6.

Volpe JJ.Neurology of the Newborn. In. Philadelphia: W.B. Sanders Company; 1995; p. 675-729.

Walker EF, Diforio D, Baum K. Developmental neuropathology and the precursors of schizophrenia. Acta Psychiatrica Scandinavica, Supplementum 1999; 395:12-9.

Watson JB, Mednick SA, Huttunen M et al. Prenatal teratogens and the development of adult mental illness. Development & Psychopathology 1999; 11:457-66.

Weinberger DR. Implications of normal brain development for the pathogenesis of schizophrenia. Archives of General Psychiatry 1987; 44:660-9.

Whitley RJ.Herpes simplex virus. In: WM Scheld; RJ Whitley; DT Durack, editors Infections of the Central Nervous System. Philadelphia: Lippencott-Raven; 1997.

Zinkernagel RM. Maternal antibodies, childhood infections, and autoimmune diseases. [see comments.]. New England Journal of Medicine 2001; 345:1331-5.

Chapter 6

MODELS AND MECHANISMS

The purpose of this chapter is to disentangle the various theories (or more appropriately, sub-theories) that have been described in previous chapters concerning the viral hypothesis for schizophrenia. In addition, the contribution of various animal models toward understanding the mechanisms proposed in these theories will be elucidated. This section is necessarily speculatively and is based on the overarching believe that both genetic and environmental factors contribute to schizophrenia. For most of the theories described, a schematic timeline depicts how environmental, genetic, and neurodevelopmental events could act in parallel, or in concert, to lead to psychopathology.

6.1 SPECIFIC MODELS

6.1.1 Infection of the Mature Brain:

The first scheme considers that schizophrenia may be caused by encephalitis in genetically-susceptible individuals. This mechanism postulates that clinical or subclinical infection near the time that psychiatric symptoms develop could act on predisposing geonotypes causing decompensation of brain function and schizophrenia psychopathology (figure 1A). This mechanism may account for only a small portion of cases and is not mutually exclusive with other environmental factors being the precipitating events acting on genetic vulnerabilities (figure 1B). Moreover, viral encephalitis could cause schizophrenia independently of predisposing genes (figure 1C); which in a sense, could be considered a "phenocopy" in that most cases would have a genetic basis but rarely a virus could produce the schizophrenia syndrome independently of a genetic anomaly.

Besides the data on human encephalitis discussed in chapter 3, there have been a plethora of studies in experimental animals, including nonhuman primates, that have clearly demonstrated that viral infections can cause a wide spectrum of measurable behavioral changes encompassing virtually every neural system (Crnic 1991; Weed et al. 2001). One advantage of such animal models is that the timing of the infection with respect to the development of symptoms is not a question, and therefore the underlying neurophysiological substrates can be more easily uncovered.

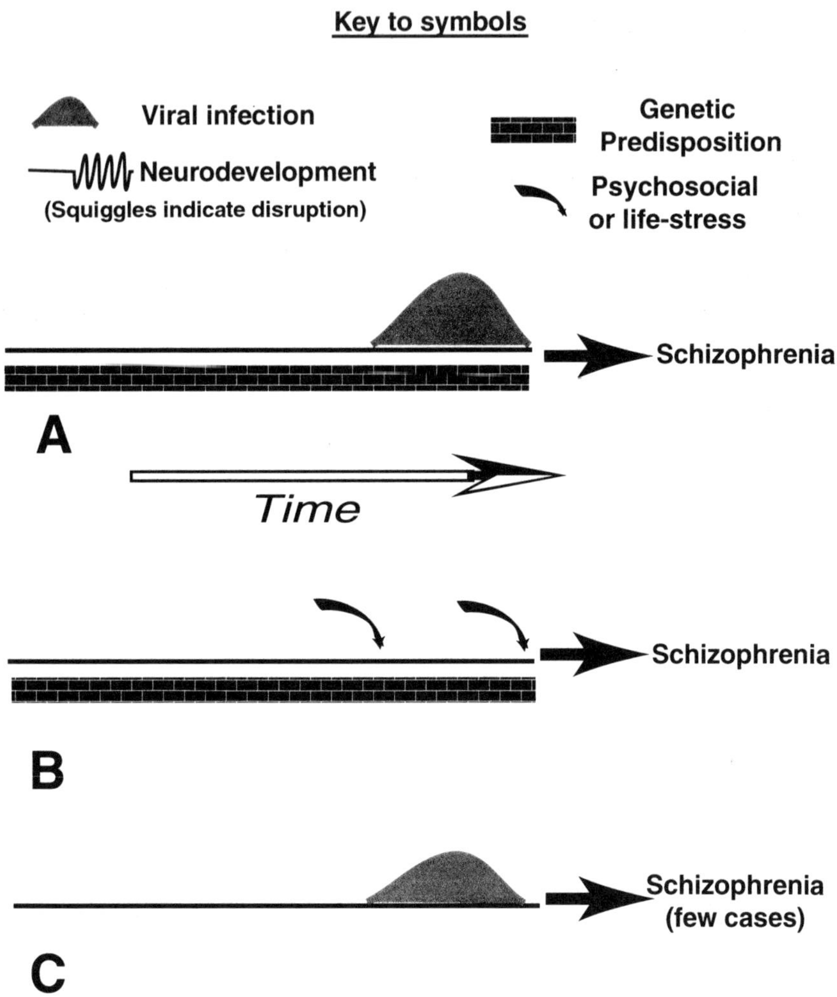

Figure 1 Timeline, beginning at conception and progressing into adulthood, showing interaction of predisposing genes and adult viral infections in the development of schizophrenia. In this figure, predisposing genotypes, which of course are present from birth, do not alter neurodevelopment but are important in mediating the impact of events later in life such as infections (A), or life stresses (B). But a genetic predisposition may not be essential for all cases of schizophrenia (C).

One obvious way that viruses could engender psychotic disorders is by altering neurotransmitters. Understandably, the belief that brain neurotransmitter abnormalities are involved in the etiopathogenesis of schizophrenia is well ensconced among psychiatric researchers and clinicians. Although this belief finds support in neurochemical studies and a massive literature in pre-clinical and clinical psychopharmacology, a unifying hypothesis linking neurotransmitter abnormalities with neurodevelopmental insults and the neuropsychological manifestations of schizophrenia has not been forthcoming, particularly with regard to the role of virus infections (Duncan et al. 1999). A prerequisite to progress in this area is gaining an understanding of how viruses can affect neurotransmitters.

The idea that viruses can influence neurotransmission is not in question. Beginning with the seminal work of Lycke and Roos, experimental infections with an array of viruses have been shown to alter CNS monoamine systems in various and complex ways (Lycke et al. 1968). While many other models have been developed subsequently, even the best animal model of schizophrenia cannot be expected to mimic all the heterogeneous behavioral and neuroanatomical features of the disease; if such a model existed it would likely be fraught with many of the inconsistencies and difficulties which impede the research of schizophrenia itself. Still, there is no doubt that viruses alter neurotransmitter function, including monoamines, glutamate and GABA. The challenge now is to determine the pathways by which these neurochemical interactions occur (Dunn et al. 1995; Watkins et al. 1995). In this regard, accumulating evidence from clinical and basic science research gives credence to the hypothesis that reciprocal interactions between cytokines produced during infections, and hormones made by the hypothalamic adrenal pituitary (HPA) axis, play a role the pathogenesis of major neuropsychiatric disorders (Benveniste 1998; Pearce 2001). Thus, viral infections are able to tap into systems known to participate in major psychiatric disorders, but their precise pathophysiological role in these diseases is not known. Interestingly, therapy with exogenous cytokines (e.g, for cancer) is well established to case a myriad of psychiatric complications including psychosis. (Capuron et al. 2001)

6.1.2 Infection During Neurodevelopment

The next theory considered proposes that a virus may be acting during development. The virus could infect the developing brain and disrupt neural morphogenesis. This could occur independently of predisposing genes. Such a mechanism may account for only a small fraction of all the schizophrenia cases (figure 2A), whereas in the majority of cases genetic factors are the main antecedents to the neurodevelopmental anomalies that are eventually brought to the surface by disease modifiers (e.g., psychosocial factors) later in life (figure 2B).

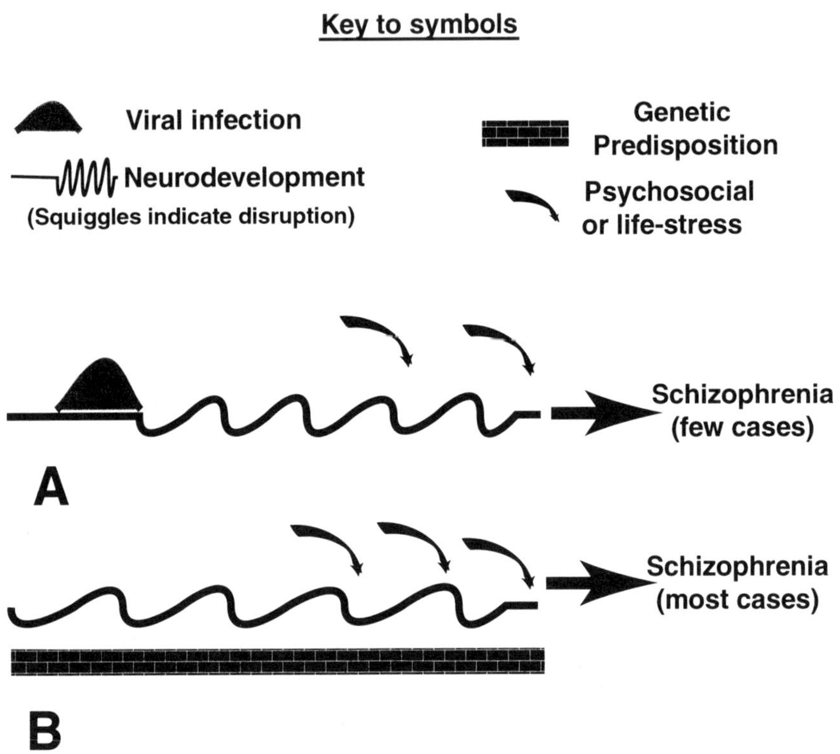

Figure 2 Timeline, beginning at conception, showing viral infections during development and their proposed relationship to schizophrenia.

Considering that there is substantial inter-individual variability among schizophrenia patients in their symptom clusters, response to treatment, and disease course, those individuals whose disease is triggered by a virus might be expected to have a qualitatively different form of schizophrenia than those whose disease is triggered by some other neurodevelopment insult. There is some tentative data to support this idea. Patients exposed to influenza during the second trimester of gestation tend to manifest prominent paranoid symptoms (Watson et al. 1999), and patients whose syndrome is dominated by deficit symptoms tend to be less likely to have winter births than the schizophrenia population at large (Kirkpatrick et al. 2001). This suggests that the type of risk factor or precursor determines schizophrenia subtypes. But the picture is far from clear, and seroprevalence for Borna diseases virus has also been associated with deficit symptoms (Waltrip et al. 1997). Thus we cannot yet conclude there is a viral form of the disease. In any case, postnatal disease modifiers such as social factors or drug abuse may help determine whether an individual develops schizophrenia and what the character of those symptoms are within the schizophrenia spectrum.

The subtype of disease that a person develops within the schizophrenia spectrum may also be determined by genetic variables or the timing in gestation of the infectious insult (Figures 3A and B). The influence of genetics are supported by the rubella study described in Chapter five (Brown et al. 2000). In that study, more than half of the patients developed a schizophrenia spectrum disorder other than schizophrenia itself, but the timing of the exposure was predominantly in the first trimester for all patients, and there were also numerous individuals who were exposed to rubella at approximately the same time in gestation who did not develop a serious mental disorder. Thus, genetics may have been the critical factor that determined whether an individual developed a psychotic disorder, as well as the character of the disorder within the schizophrenia spectrum. However, Machon et al. have found no evidence that a genetic predisposition determines a schizophrenia diagnosis after second trimester exposure to influenza (Machon et al. 1995).

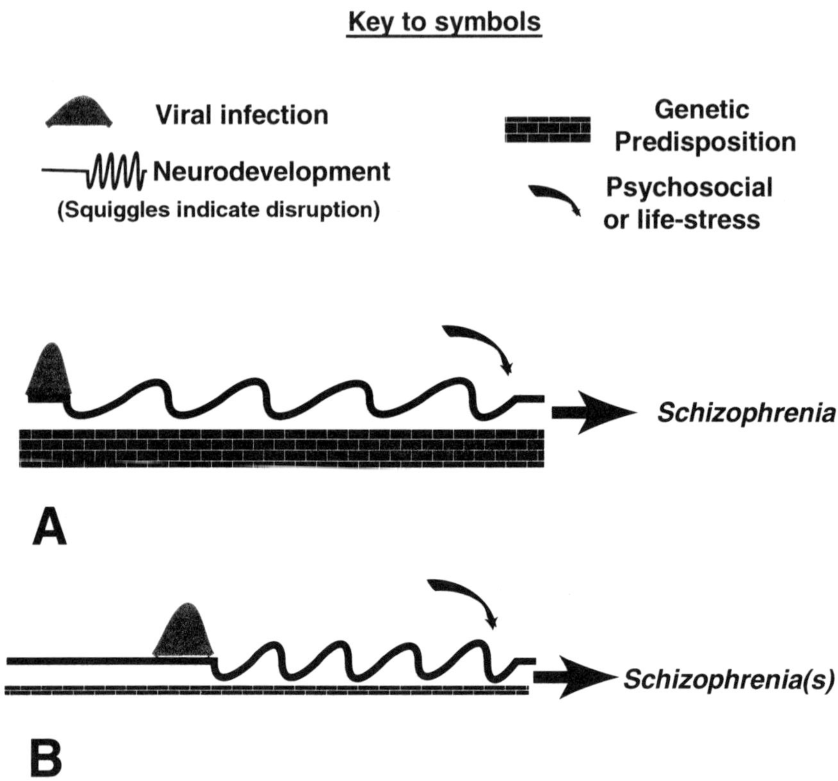

Figure 3 Timeline showing proposed influence of genetic loading and timing of infection on schizophrenia phenotypes,

Animal models have supported both genetic factors and timing of the viral infection during development as being important variables in neurobehavioral sequela (Engel et al. 2000; Hornig et al. 1999; Ludwig et al. 1997; Pearce 2000; Richt et al. 1997; Rubin et al. 1999). Borna diseases virus has been extensively studied in this regard, and the impact of infection on an array of behavioral parameters has been shown to be dependent on the genetic background and developmental stage of the experimental animal being examined. Neonatal infection of rodents with Borna disease virus has proven to be an interesting model for neuropsychiatric disorders because it elicits a variety of alterations in complex behaviors including learning, motor development, emotional response and social interactions. One of the strengths of the Borna models is that the pathophysiology in animals may mirror some aspects of the infection in humans. The potential connection of BDV with schizophrenia is discussed in chapter 3.

One advantage of using rodent models in general is that the development of some brain regions (e.g. the hippocampus) occurs neonatally and corresponds to the development of the human in the second trimester in utero. Therefore early postnatal infections, which can be achieved more consistently than prenatal infections in an experimental setting, can be used to recapitulate the interaction of viruses with the developing human nervous system in utero. We have used infection with lymphocytic choriomeningitis virus (LCMV) in neonatal rats to examine the long-term sequela of a viral infection during neurodevelopment on the structure and function of the hippocampus-- a brain region known to be affected in schizophrenia.

Like Borna disease virus, LCMV neuropathology is dependent on the genetic background of the host and the developmental stage of the brain at the time of infection. Although LCMV is non-cytopathic to neurons, the virus is capable of causing latent hippocampal abnormalities that appear to progress even after infectious virus is cleared (Pearce 2000). Hippocampal dentate granule cells and GABAergic interneurons are infected early in the disease. The dentate granule cells are mitotically active at the time of the infection, and while GABAergic neurons are thought to cease dividing prenatally, they are undergoing the final stages of differentiation during this postnatal interval. GABAergic neurons appear to become impaired by the infection. This finding is based in part on electrophysiological recordings in the dentate gyrus of adult rats, originally infected with LCMV as neonates; these rats have a marked defect in inhibitory neurotransmission. Accordingly, immunohistochemistry has revealed a decrease in a subset of GABAergic cells (those expressing parvalbumin). As the rats proceed toward sexual maturity, degeneration of dentate granule cells becomes apparent, and the adults manifest varying degrees of hippocampal volume reduction and lateral ventricular dilation.

Because GABAergic neurons regulate the amount of excitatory input to dentate granule cells (principle cells), these principle cells receive unchecked excitatory (glutamatergic) input in the infected animals. Thus neonatal LCMV infection appears to cause disinhibition of principle cells within the hippocampal dentate gyrus. This deficient inhibitory control could trigger a pathophysiological cascade in which glutamatergic input may be released from its inhibitory control thus leading to the latent death of dentate granule cells as a consequence of synaptically-mediated excitotoxicity (Pearce 2000; Pearce et al. 2000). Other viruses could lead to analogous alterations in the balance between inhibition and excitation depending on their tropisms and the time of infection, and principle cells other than dentate granule neurons could consequently be lost.

The importance of the LCMV model is that it suggests a new mechanism that could establish a pathophysiological nexus between a viral infection and some recent enigmatic findings concerning the neurochemistry and disease course in schizophrenia. Specifically, I propose that some cases of schizophrenia could be due to "Virus-Induced Disinhibition and Excitotoxicity" (VIDE). The VIDE mechanism invokes a specific pathophysiological connection between a perinatal viral infection, and the disease course and terminal neuropathological features of schizophrenia. Many of the parallels predicted by this mechanism have been borne-out in recent human experiments. These have been explained in detail elsewhere are only mentioned briefly here (Pearce 2000).

Specifically, a viral infection of the developing pre-schizophrenic brain is posited to occur during cellular differentiation of corticohippocampal GABAergic interneurons. Assuming the causative virus is noncytopathic like LCMV, infected GABAergic interneurons would survive the initial infection while still undergoing disruption of non-essential (luxury) cell functions. Early impairment of inhibitory neurotransmission might then give rise to motor and cognitive disturbances that often presage schizophrenia (Walker et al. 1994). The VIDE mechanism would predict that some indices of GABAergic neurons would be decreased at autopsy, and that parvalbumin neurons would be found specifically-lacking in the hippocampus. Recent autopsy studies support both these predictions (Benes 2000; Zhi et al. 2002). Moreover, we have suggested that the loss of parvalbumin neurons may result from a cytokine-mediated disruption of brain-derived neurotrophic factor on which these neurons depend for their normal differentiation (Pearce 2000). This links the VIDE mechanism to the immune-mediated theory described below. And finally, VIDE is predictive of heterogeneity in both psychiatric symptom severity and neurodegeneration (Woods 1998). Specifically, this concept is based on the likelihood of individual differences in the reserve capacity of inhibitory neuronal networks that can be brought to bear in a homeostatic effort to override the virus-induced disinhibition, and the subsequent excitotoxicity. Thus, individuals with inherently more robust inhibitory circuits, or resistance to the disruption of those circuits by the infection, would have more mild psychiatric symptoms, and presumably less degeneration. However, psychiatric and degenerative components of the disease could diverge because other compensatory mechanism would obviously come into play during the excitotoxicity stage.

Using another rodent model, Fatemi et al. have examined the impact of prenatal influenza infection on the development of the hippocampus (Fatemi et al. 1999). Infection of mice at mid gestation was found to cause a significant reduction in Reelin in cortical and hippocampal brain regions when the mice were examined postnatally. This is an important experiment because it utilized a virus that has been implicated in schizophrenia, and alterations in Reelin (a protein involved in normal brain development) have been found in schizophrenic brain. An additional strength of this model is demonstrated by studies showing that the offspring of influenza-infected mice display deficits in pre-pulse inhibition in the acoustical startle response (Patterson 2002). Importantly, deficits in pre-pulse inhibition have been extensively described in schizophrenia, and this behavioral test represents one of the most consistent phenomena that can be used to develop antipsychotic drugs. Abnormalities in pre-pulse inhibition have also been found after neonatal infection with herpes simplex virus or cytomegalovirus in rats.

6.1.3 Virus-induced Immune Molecules During Development

Another possible mechanism by which the viral infection during gestation could cause developmental irregularities leading to schizophrenia is by the induction of immune molecules involved in brain development (Figure 4A). These immune molecules include (but are not limited to) antibodies and cytokines (Patterson 2002; Wright et al. 1993). There is burgeoning interest in the role of cytokines in CNS ontogeny. Numerous cytokines produced by immune cells are also expressed by neural cells in a developmentally-regulated manner. Several proinflammatory cytokines such as IL-1, IL-6, and TNF-alpha have been shown to influence the growth and differentiation of neural cells. Indeed, it is becoming increasingly apparent that there is no consistent line of demarcation between cytokines responsible for regulating immune function, and growth or differentiation factors responsible for governing CNS development. The importance of cytokines in human brain disorders is reflected in a growing literature that shows that in pregnancies complicated by chorioamnionitis, elevations in cytokines measured in maternal amniotic fluid correlate with CNS anomalies in the neonate (Dammann et al. 2000; Foster-Barber et al. 2002).

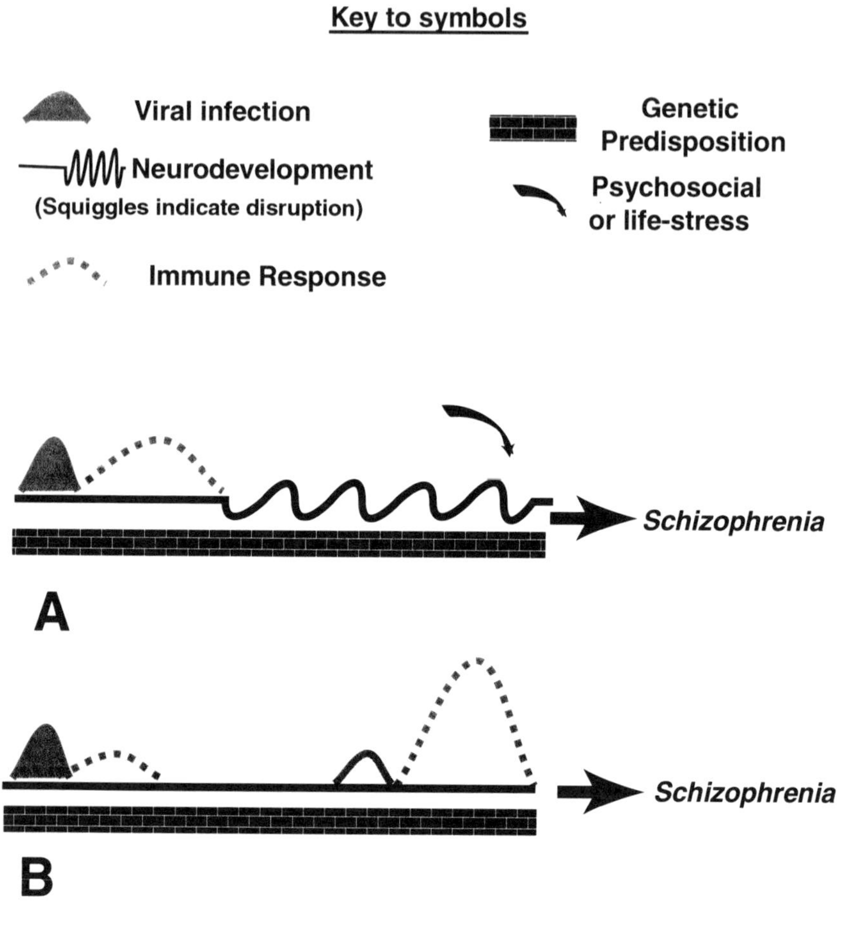

Figure 4 Scheme for the role of virus-induced immune molecules in schizophrenia pathogenesis.

A variation on the developmental-immune theory suggests that an environmental event in adolescence such as infection with another virus could trigger antibodies that cross-reacted with brain constituents (Figure 4B). An earlier viral infection that primed this secondary immune response could also be postulated to have occurred, but is not essential to the theory. On the other hand, the proximal event occurring in adolescence would not necessarily have to be an infection. Any number of the immunomodulatory processes, including stress, which induces the hormone cortisol, could serve as the proximal event by regulating the immune response.

Immune molecules must traverse the blood brain barrier in order to affect neurodevelopment or cause functional alterations of the adult brain. A critical role for the BBB in schizophrenia must be considered because this structure sequesters the CNS from invading pathogens as well as deleterious components of the immune response; and because individual differences in BBB development are presumably determined by genetic, as well as maternal factors such as intrauterine hypoxia (Kirch 1993).

Functionally, the BBB is a complex structure comprised of multiple cell types, and its ability to exclude infectious agents and pernicious immune influences is poorly understood. A delay in a crucial aspect of BBB development could leave the immature brain vulnerable to a group pathogens having the capacity to provoke a common cassette of host responses such as the elaboration of proinflammatory cytokines. These host molecules might disrupt neurodevelopment only in fetal brains unshielded by the BBB, thus leading to pre-schizophrenic neuro-circuitry alterations. Interestingly, a substantial portion (7-33%) of schizophrenia patients demonstrate evidence of BBB abnormalities. These alterations do not appear to be a consequence of antipsychotic medication. Nonetheless, according to neurodevelopmental theories, a defect in the BBB would be expected to precede schizophrenia. Since there are likely multiple pathophysiological routes leading to BBB impairment, the minority of patients actively manifesting BBB abnormalities after diagnosis may have succumbed to one of the pathways in which the developmental abnormality persists into adulthood. Like many putative developmental antecedents of schizophrenia, early defects in the BBB are difficult to study in humans but remain important because they could represent a crucial inherent component in the multi-factorial genetic: environmental model of schizophrenia.

6.1.4 Infection and Pregnancy Complications

A multitude of labor and delivery complications have been identified as risk factors in the subsequent development of schizophrenia including prolonged labor, abnormal fetal presentation, premature rupture of membranes, maternal hemorrhage, and infective chorioamnionitis (Verdoux et al. 1997). Infections are well known to be instigators of obstetric complications including prematurity. Thus, it's possible that obstetric complications are the critical factor leading to neurodevelopmental disruptions (e.g. via common mechanisms such as fetal hypoxia) and that infections are merely one of the many causes of these complications (Figure 5). I believe this mechanism has merit, but it is also clear that obstetric complications are not sufficient on their own to cause schizophrenia because the rate of obstetric complications in any relevant population far exceeds that of schizophrenia. Thus, one must argue for the importance of a genetic vulnerability on the part of the fetus. Nevertheless, Machon et al. have argued against a mechanism involving virus-induced pregnancy complications based on their sound (albeit preliminary) data (Machon et al. 1995). Specifically, in their epidemiological investigations of second trimester exposure to influenza, the patients who developed schizophrenia actually evidenced less obstetric complications than controls.

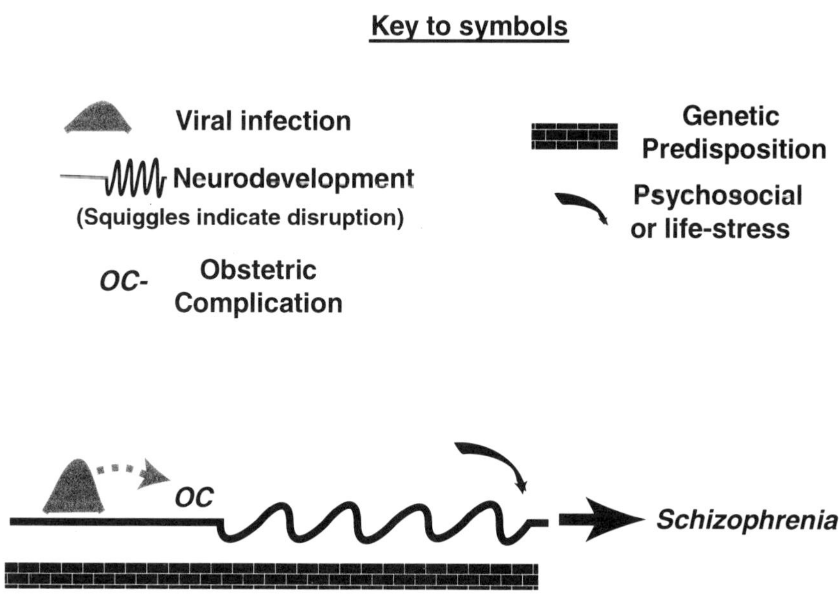

Figure 5 Scheme showing viral infections as instigators of obstetric complications.

A variation on the infectious-obstetric complications theory can be considered as well. There may be common soluble biological response modifiers that are induced by infections as well as other adverse events during pregnancy. Thus infections and stress during pregnancy might both lead to the production of the same deleterious hormone or cytokine. Our laboratory is currently investigating this hypothesis.

6.1.5 Retroviral Paradigms

The idea that a retrovirus could be involved in schizophrenia was first proposed in 1984, and since that time the hypothesis has been refined and subjected to preliminary testing (Crow 1984; Deb-Rinker et al. 1999; Hart et al. 1999; Karlsson et al. 2001; O'Reilly et al. 1996). The basic structure of the retrovirus virion is not notably different from other viral particles. However, the genetic "payload" of the virus is a double stranded RNA molecule that by virtue of the enzyme reverse transcriptase is transcribed into DNA. The DNA in turn becomes integrated into the chromosomal DNA of the host cell where it becomes a provirus, permanently altering the genotype the host cell and all of its progeny. Thus retroviruses can change the genome of the host.

There are four retrovirus that are known to cause disease in humans: HTLV-1, HTLV-II, HIV-2, and HIV-2. The encephalitis caused by HIV has already been discussed. HTLV-1 causes the neurological disease, tropical spastic paraparesis, in a small portion of people infected. Recent data implicate immunogenetic factors as partial determinants of which infected patients develop this neurological complication (Usuku et al. 1988). Thus far, the pathophysiological properties of the known human retroviruses do strongly implicate them as causes of schizophrenia.

Based on the unique properties of retroviruses, there are several conceivable mechanisms by which a presumably unidentified viral species could cause schizophrenia. For the sake of brevity, I will only consider those theories that are contingent on the ability of these viruses to integrate into host DNA and consequently precipitate pathophysiological events beyond those discussed above for other viruses. Because retroviruses integrate into the existing host genome, they can cause insertional mutagenesis, consequently altering the regulation of nearby genes. In theory, a retrovirus could attack the human brain during adulthood and disrupt the function of normal genes (e.g. by integrating into the regulatory region of the D-2 dopamine receptor gene). The resulting perturbation in brain activity would then presumably lead to psychotic symptoms (Figure 6A)

A retrovirus might instead integrate near genes controlling neurodevelopment, consequently disturbing the developmental program and leaving the affected individuals vulnerable to later disease modifiers such as psychosocial stress (Figure 6B). The earlier in embryogenesis that the insertion of the provirus took place, the more cells would be affected. But the pathophysiological features of the developmental irregularities would depend on when and where the affected gene was needed during neuromorphogenesis. Retroviruses were first proposed to integrate into a site regulating hemispheric cerebral development (Crow 1984). Although there is little direct data to support this specific mechanism, there is a large body of evidence supporting the notion that schizophrenia is associated with structural and functional abnormalities in cerebral lateralization. The potential role of viruses in this process has been relatively neglected. Even non-retroviruses could cause disturbances in the laterality of brain development by tracking along nerve fibers and entering only one side of the brain, or through immune mechanisms.

If a retrovirus integrated into the germ line it could be transmitted vertically from either the mother or father to the schizophrenia patient. Then it could exert its pernicious influence during development, as described above, or it could lie dormant until reactivated to produce new virions. Such reactivation is (in a matter of speaking) under host control. Steroid hormones are among the prime candidates that could be physiological mediators of such reactivation. Specifically, sex hormones have long been hypothesized to play a role in triggering schizophrenia because the age of disease onset corresponds with adolescence or early adulthood when these hormone levels are changing. In addition, the adrenal hormone, cortisol, has recently been shown to be altered during the schizotypal "prodrome" of schizophrenia (Weinstein et al. 1999). Retroviruses are known to be under regulatory control by host steroid hormones. For example, mouse mammary tumor virus is a retrovirus that can be transmitted (in mice) horizontally, or vertically through the germ line. Transcription of the mouse mammary tumor virus genome is strongly induced by cortisol (or the mouse equivalent, corticosterone).

Since cortisol is a stress-activated hormone, adverse life events could cause reactivation of an integrated provirus. The new virions produced could cause cytopathic damage to neurons and could disseminate to other brain regions leading to a degenerative disease course. The cyclical activity that is inherent in the retrovirus life cycle, and is modulated by host factors, could lead to a relapsing-remitting psychopathology.

The human genome of every person has retrovirus sequences imbedded within it. On the whole, these do not represent de novo infections, but are stretches of retroviral DNA that have been retained over the course of vertebrate evolution. While many of these are incomplete and do not contain opened reading frames, some proviral sequences capable of making viral proteins have been identified. Interestingly, the placenta appears to be a preferred site of expression.

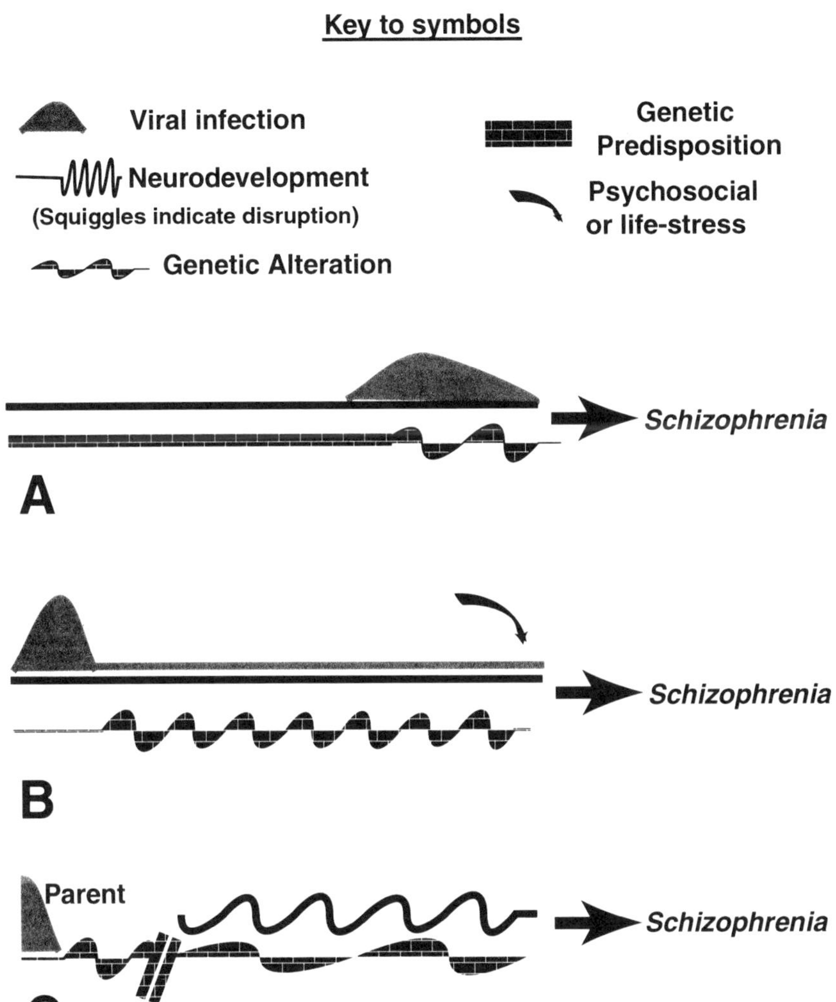

Figure 6 Timeline for the impact of retroviral infections on host genotype. In most cases only a subset of cells will be genetically-altered by the virus. In figure 6 C the parent transmits the retrovirus to the child.

Considering that all humans have these so-called endogenous retroviruses, it's possible that the transcription of these sequences could be activated by a hormone, or by immune molecules engendered by a bacterial or viral infection (Figure 7). Expression of these endogenous retroviruses could disrupt development in a manner analogous to that shown in the figure 6B, or could be closely related to the immediate brain alterations occurring proximal to psychotic symptoms as depicted in figure 6A. Karlsson et al. have provided evidence for the latter. They found endogenous retroviral RNA sequences in the CSF of 10 out of 35 patients (29%) with recent onset schizophrenia or schizoaffective disorder, but only in 5% of chronic patients, and none of 30 controls (Karlsson et al. 2001).

If retroviral RNA can be translated into antigenic proteins, as has been suggested to occur in other human diseases such as insulin dependent diabetes mellitus, then the reactivation of endogenous retroviruses may provoke an immune response against brain antigens; this concept therefore buttresses the retroviral theory of schizophrenia with the autoimmune data.

The potential role of retroviruses opens an exciting new phase in the research of the viral hypothesis of schizophrenia, and illustrates the dependence of this scientific inquiry on advances in virology and molecular biology. When considering the various theories described above, it is important to realize that while most are probably wrong, they are subject to refinement only though proper experimentation. Such experimentation is essential considering that viral diseases are preventable and increasingly treatable, and if only a fraction of schizophrenia cases were caused by viruses, their prevention would still be a worthy goal.

6.2 A HISTORY LESSON FROM ONCOLOGY CONCERNING THE VIRAL HYPOTHESIS

The belief that cancer could be caused by a contagious agent goes back centuries, but the idea was revived briefly in the early 1900's as more and more infectious agents were being discovered to cause idiopathic diseases. Nonetheless, interest soon subsided, and for most of the mid 1900's the now famous experiment by Peyton Rous showing that a virus can be carcinogenic in chickens was relegated to birds. Rous was discouraged by the ridicule of his ideas and experiments, and he stopped working on the project while he was still a relatively young man (Epstein 2001).

Any progress in showing the role of viruses in human cancer was constantly shadowed by stringent criticism throughout most of the 20th century. Even in the late 1970's and early 1980's, viral oncogenesis in humans remained highly controversial.

The reality that cancer has a number of different causes, including viruses, has finally nudged its way into scientific dogma. Gradually, it has become accepted that there are multiple ways that viruses may participate in the causal concatenation of events leading to the development of a neoplasm. Most experts in oncology now accept that viruses are responsible, at some level in the casual chain, for about 20 percent of all human cancers (Epstein 2001).

Many experts in schizophrenia have been calling for the recognition that there is no one disease called schizophrenia, but rather a syndrome with a heterogeneous symptom complex that may be underlaid by different etiologies (Kirkpatrick et al. 2001). If we take our lesson from other diseases characterized by complex pathophysiology and multiple risk factors, viruses warrant our greatest attention as one of those etiologies.

6.3 WHAT A VIRAL ETIOLOGY OF SCHIZOPHRENIA DOES NOT MEAN

I can envision only one extremely deleterious consequences of accepting a viral etiology for schizophrenia: avoiding contact with patients for fear of being infected. There is absolutely no data to suggest that patients with schizophrenia are actively contagious with a virus that could cause mental illness in an unaffected individual. Even the scant data on Borna disease virus suggesting that some mental-health care workers may have acquired the virus in a hospital setting suggests that the virus has only an adventitious connection with mental illness. All the available data indicate that ifa virus causes schizophrenia it is either long gone or not contagious by the time psychiatric symptoms arise.

6.4 CONCLUSION

Schizophrenia is a devastating condition of the human mind. When the disease is at its worst, the patient--the human being--is under assault by his own brain; which according to the viral hypothesis was previously assaulted by a virus. If a virus causes schizophrenia the options for prevention and treatment are profoundly expanded. For that reason (if no other) there should be a growing sense of obligation among those of us who study the disease to determine the veracity of the viral hypothesis.

REFERENCES

Benes FM. Emerging principles of altered neural circuitry in schizophrenia. Brain Research - Brain Research Reviews 2000; 31:251-69.

Benveniste EN. Cytokine actions in the central nervous system. Cytokine & Growth Factor Reviews 1998; 9:259-75.

Brown AS, Cohen P, Greenwald S et al. Nonaffective psychosis after prenatal exposure to rubella. American Journal of Psychiatry 2000; 157:438-43.

Capuron L, Bluthe, acute RM et al. Cytokines in clinical psychiatry [9]. American Journal of Psychiatry 2001; 158:1163-4.

Crnic LS.Behavioral consequences of virus infections. In: R Ader; DL Felton; N Cohen, editors Psychoneuroimmunology. 2nd edn. San Diego: Academic Press; 1991; p. 749-68.

Crow TJ. A re-evaluation of the viral hypothesis: is psychosis the result of retroviral integration at a site close to the cerebral dominance gene? British Journal of Psychiatry 1984; 145:243-53.

Dammann O, Leviton A. Role of the fetus in perinatal infection and neonatal brain damage. Current Opinion in Pediatrics 2000; 12:99-104.

Deb-Rinker P, Klempan TA, O'Reilly RL et al. Molecular characterization of a MSRV-like sequence identified by RDA from monozygotic twin pairs discordant for schizophrenia. Genomics 1999; 61:133-44.

Duncan GE, Sheitman BB, Lieberman JA. An integrated view of pathophysiological models of schizophrenia. Brain Research - Brain Research Reviews 1999; 29:250-64.

Dunn AJ, Wang J. Cytokine effects on CNS biogenic amines. Neuroimmunomodulation 1995; 2:319-28.

Engel JA, Zhang J, Bergstrom T et al. Neonatal herpes simplex virus type 1 brain infection affects the development of sensorimotor gating in rats. Brain Research 2000; 863:233-40.

Epstein MA. Historical background. Philosophical Transactions of the Royal Society of London - Series B: Biological Sciences 2001; 356:413-20.

Fatemi SH, Emamian ES, Kist D et al. Defective corticogenesis and reduction in Reelin immunoreactivity in cortex and hippocampus of prenatally infected neonatal mice. Molecular Psychiatry 1999; 4:145-54.

Foster-Barber A, Ferriero DM. Neonatal encephalopathy in the term infant: neuroimaging and inflammatory cytokines. Mental Retardation & Developmental Disabilities Research Reviews 2002; 8:20-4.

Hart DJ, Heath RG, Sautter FJ, Jr. et al. Antiretroviral antibodies: implications for schizophrenia, schizophrenia spectrum disorders, and bipolar disorder. Biological Psychiatry 1999; 45:704-14.

Hornig M, Weissenbock H, Horscroft N et al. An infection-based model of neurodevelopmental damage. Proceedings of the National Academy of Sciences of the United States of America 1999; 96:12102-7.

Karlsson H, Bachmann S, Schroder J et al. Retroviral RNA identified in the cerebrospinal fluids and brains of individuals with schizophrenia. [see comments.]. Proceedings of the National Academy of Sciences of the United States of America 2001; 98:4634-9.

Kirch DG. Infection and autoimmunity as etiologic factors in schizophrenia: a review and reappraisal. Schizophrenia Bulletin 1993; 19:355-70.

Kirkpatrick B, Buchanan RW, Ross DE et al. A separate disease within the syndrome of schizophrenia. Archives of General Psychiatry 2001; 58:165-71.

Ludwig H, Bode L. The neuropathogenesis of Borna disease virus infections. Intervirology 1997; 40:185-97.

Lycke E, Roos BE. Effect on the monoamine-metabolism of the mouse brain by experimental Herpes simplex infection. Experientia 1968; 24:687-9.

Machon RA, Mednick SA, Huttunen MO.Fetal viral infection and adult schizophrenia:empirical findings and interpretation. In: SA Mednick; JM Hollister, editors Neural development and schizophrenia. New York: Plenum; 1995.

O'Reilly RL, Singh SM. Retroviruses and schizophrenia revisited. American Journal of Medical Genetics 1996; 67:19-24.

Patterson PH. Maternal infection: window on neuroimmune interactions in fetal brain development and mental illness. Current Opinion in Neurobiology 2002; 12:115-8.

Pearce BD. Neuroendocrine-immune interactions during viral infections. Advances in Virus Research 2001; 56:465-509.

Pearce BD.Viral mechanisms of schizophrenia. In: M Myslobodsky; I Weiner, editors Contemporary Issues in Modeling Psychopathology. Norwell: Kluwer Academic; 2000; p. 79-102.

Pearce BD, Valadi NM, Po CL et al. Viral infection of developing GABAergic neurons in a model of hippocampal disinhibition. NeuroReport 2000; 11:2433-9.

Richt JA, Pfeuffer I, Christ M et al. Borna disease virus infection in animals and humans. Emerging Infectious Diseases 1997; 3:343-52.

Rubin SA, Bautista JR, Moran TH et al. Viral teratogenesis: brain developmental damage associated with maturation state at time of infection. Brain Research. Developmental Brain Research 1999; 112:237-44.

Usuku K, Sonoda S, Osame M et al. HLA haplotype-linked high immune responsiveness against HTLV-I in HTLV-I-associated myelopathy: comparison with adult T-cell leukemia/lymphoma. Annals of Neurology 1988; 23 Suppl:S143-50.

Verdoux H, Geddes JR, Takei N et al. Obstetric complications and age at onset in schizophrenia: an international collaborative meta-analysis of individual patient data [see comments]. American Journal of Psychiatry 1997; 154:1220-7.

Walker EF, Savoie T, Davis D. Neuromotor precursors of schizophrenia. Schizophrenia Bulletin 1994; 20:441-51.

Waltrip RW, 2nd, Buchanan RW, Carpenter WT, Jr. et al. Borna disease virus antibodies and the deficit syndrome of schizophrenia. Schizophrenia Research 1997; 23:253-7.

Watkins LR, Maier SF, Goehler LE. Cytokine-to-brain communication: a review and analysis of alternative mechanisms. Life Sciences 1995; 57:1011-26.

Watson JB, Mednick SA, Huttunen M et al. Prenatal teratogens and the development of adult mental illness. Development & Psychopathology 1999; 11:457-66.

Weed MR, Gold LH. Paradigms for behavioral assessment of viral pathogenesis. Advances in Virus Research 2001; 56:583-626.

Weinstein DD, Diforio D, Schiffman J et al. Minor physical anomalies, dermatoglyphic asymmetries, and cortisol levels in adolescents with schizotypal personality disorder. American Journal of Psychiatry 1999; 156:617-23.

Woods BT. Is schizophrenia a progressive neurodevelopmental disorder? Toward a unitary pathogenetic mechanism [see comments]. American Journal of Psychiatry 1998; 155:1661-70.

Wright P, Gill M, Murray RM. Schizophrenia: genetics and the maternal immune response to viral infection. American Journal of Medical Genetics 1993; 48:40-6.

Zhi JZ, Reynolds GP. A selective decrease in the relative density of parvalbumin-immunoreactive neurons in the hippocampus in schizophrenia. Schizophrenia Research 2002; 55:1-10.

INDEX